William Taylor Adams

Stem to Stern

Building the Boat

William Taylor Adams

Stem to Stern
Building the Boat

ISBN/EAN: 9783337413118

Printed in Europe, USA, Canada, Australia, Japan

Cover: Foto ©berggeist007 / pixelio.de

More available books at **www.hansebooks.com**

THE BOAT-BUILDER SERIES

STEM TO STERN

OR

BUILDING THE BOAT

OLIVER OPTIC.

AUTHOR OF "YOUNG AMERICA ABROAD," "THE GREAT WESTERN SERIES," "THE
ARMY AND NAVY SERIES," "THE WOODVILLE SERIES," "THE STARRY-
FLAG SERIES," "THE BOAT-CLUB STORIES," "THE ONWARD AND
UPWARD SERIES," "THE YACHT-CLUB SERIES," "THE
LAKE-SHORE SERIES," "THE RIVERDALE SE-
RIES," "ALL ADRIFT," "SNUG HARBOR,"
"SQUARE AND COMPASSES,"
ETC., ETC.

With Illustrations

BOSTON

LEE AND SHEPARD PUBLISHERS

NEW YORK CHARLES T. DILLINGHAM

1886

TO

The Boat-Builder Series.

PREFACE.

"Stem to Stern" is the fourth volume of the
"Boat-Builder Series." Most of the characters
connected with the Beech Hill Industrial School
continue to take part in the action of the story.
Like its predecessors, a considerable portion of
the work is devoted to business and mechanical
information. The writer finds it quite impracti-
cable to give as minute directions for the building
of a boat as a few of his young readers may desire, for the entire volume would hardly afford
sufficient space for all the details of planning and
constructing a yacht. But he has endeavored to
impart some information in a general way in regard
to shipbuilding, and has indicated in what manner
the ambitious young boat-builder may obtain the
amplest instruction in this difficult art. It is
necessary to assure his young friends that, with all

5

the book-knowledge it is possible to obtain on the
subject, it will require a great deal of skill and not
a little scientific and technical learning to enable
him to construct anything more elaborate than
an ordinary flatboat. Nothing but assiduous practice can procure the skill, and nothing but hard
study the geometrical and technical details of the
art.

As in the preceding volumes of the series,
"STEM TO STERN" is largely a story of adventure
on Lake Champlain and its shores. A new character is introduced as the leading spirit of the
story, whose struggles with the difficulties in his
life-path can hardly fail to interest the young
reader. Though he is peaceful and submissive
under ordinary circumstances, with none of the
swellish importance of many boys of his years,
he is not a milk-and-water youth, and has pluck
and strength enough to "stand up" for those whom
misfortune has placed under his protection.

Although the two remaining volumes of the series
are especially devoted to rigging and sailing a boat,
the present and the preceding books incidentally
treat of these subjects. While so many young men

on the sea, lakes and rivers seem to inherit or early acquire a taste for boats and boating, it is important that they should understand the theory of managing a sailing craft, though nothing but intelligent practice can make a competent "skipper." With such knowledge and skill, boat-sailing is a safe, as well as a healthy and improving sport.

As in former volumes, the writer has endeavored to interest his young readers in mechanical operations and pursuits; and he hopes the series will contribute its mite in influencing boys to respect manual labor and to adopt it as a pastime or the business of life.

DORCHESTER, MASS., August 17, 1885.

CONTENTS.

9

CHAPTER VIII.

CHAPTER IX.

CHAPTER X.

CHAPTER XI.

CHAPTER XII.

CHAPTER XIII.

CHAPTER XIV.

CHAPTER XV.

CHAPTER XVI.

CHAPTER XVII.

CHAPTER XVIII.

CHAPTER XIX.

CHAPTER XX.

CHAPTER XXI.

CHAPTER XXII.

CHAPTER XXIII.

CHAPTER XXIV.

CHAPTER XXV.

CHAPTER XXVI.

CHAPTER XXVII.

CHAPTER XXVIII.

CHAPTER XXIX.

CHAPTER XXX.

STEM TO STERN;

OR,

BUILDING THE BOAT.

CHAPTER I.

LILY BRISTOL AND HER TORMENTOR.

I DON'T want anybody to row for me, Mr.
Walker; I came out to take a little exercise,
and I can do it best when I am all alone," said
Miss Lily Bristol to a young gentleman of about
eighteen who stood on the sandy beach.

"But it will be a good deal more sociable to
have company," replied Walk Billcord with a
smile and a smirk.

Lily Bristol had the reputation of being a very
pretty girl, and fame had not exaggerated her
beauty. She was very plainly dressed, but she
was as neat as though she had just come out of the
bureau-drawer. She was seated in a rude flatboat,
with a pair of oars in her hands, which she seemed
to know how to use.

13

The boat was only a rod or two from the end of Sandy Point, at the southern side of the entrance to a bay with the same name. It was in the spring of the year, and the water in Lake Champlain was at its highest.

Hardly more than a rod from the point where the rippling waves sported with the bright sand was a small and lightly-built cottage. It contained two rooms on the lower floor, with two small attic chambers over them. The structure rested on posts set in the sand, and looked as light and airy as a bird-cage.

This cottage was the home of Peter Bristol, or, rather, of his wife and two children; for the father of the family had been away for two years, seeking to better his impaired fortunes. Peter had always been a poor man, and was always likely to be. He had been a sort of Jack-at-all-trades, not particularly good at any. He had been a fireman on a railroad, a farm-hand, a general jobber; he had tried his hand at almost everything without much success.

Major Billcord owned all the land near Sandy Point. Some years before, he had taken it into his head that the high ground in the rear of Sandy Bay would be an excellent site for a hotel. Some of his friends did not agree with him, and assured him that a hotel could not live in this location.

But the major was an obstinate man, and had his own way. He erected a structure of fifty rooms, with the intention of adding a hundred more after the first season. But for half a dozen reasons the hotel was a dreary failure. It never contained more than half a score of guests at any one time.

Included in this small number was Colonel Buckmill, who was then looking for a suitable site for an academy. The owner of the estate would not admit that the hotel was a failure, but he hinted that the building might be obtained for the school. It exactly suited Colonel Buckmill, and a bargain was soon made for a lease of it. In this manner the Sunnyside Hotel became the Chesterfield Collegiate Institute in the autumn of the same year.

Of course, one of the attractions of the Sunnyside was to be boating on the lake, and Major Billcord provided two sailboats and some rowboats; and Peter Bristol, who was a good boatman, was engaged to take care of the boats, and act as skipper when required. The poor man, taking his cue from his employer, believed he had fallen upon a bonanza. His fortune was made, and the rest of his days would be spent at Sandy Point.

His wife had over three hundred dollars in her own right in a savings bank, which she was willing to put into a house, and the cottage on the point was built. The family moved into it, and were delighted with the situation, though it was a rather dismal place in the winter. Peter was to have half the money derived from letting the boats; but he soon found that he had nothing to do. The few guests did not care to row or sail.

The boatman had no rent to pay, for the major had given him permission to put his house on the point without charge; but he found it was very hard work to get enough for his family to eat. Lily obtained work in Westport, and Paul attended to the boats while his father worked at haying, and they got through the season. But the dream of fortune had collapsed.

Peter Bristol was discouraged, and went to New York to find work. He obtained no situation, and shipped for the West Indies. A letter from him informed his family that he was at work on a plantation, and he hoped to do well after a while. Since that, nothing had been heard from him in two years.

Paul obtained a little work at the institute, and Lily kept her place in Westport; so that the family had worried along until the daughter lost her situation for the want of sufficient work at the store in which she was employed. Then it was difficult even to obtain enough to eat. Paul did his best, and allowed himself to be bullied and kicked by the gentlemanly students of the institute, while he could make an occasional quarter.

Major Billcord lived in Westport, and his son had lately become a pupil in the institute. He was older than most of the students, and was a wild young fellow. In the early spring he had seen Lily Bristol. He agreed with others who had seen her that she was a remarkably pretty girl, and he had made frequent visits to Sandy Point.

"I prefer to be in the boat alone," Lily replied to the young gentleman's remark that it would be more sociable to have company.

"But I want to see you, Lily, and have a talk with you," persisted Walker Billcord.

"I will see you at the cottage if you desire," answered Lily.

"But I wish to see you alone."

"You cannot see me alone, sir," replied the pretty maiden with a great deal of spirit.

"What's the reason I can't? I shall not hurt

you. I think I know how to behave like a gentleman."

"Perhaps you do," added Lily rather doubt-
fully, for Walk Billcord's reputation was none of
the best.

"If you will come to the shore, I will row you
all about the bay," Walk insisted. "I will make
it as pleasant for you as possible."

"No, I thank you," replied the damsel decidedly.

"What's the matter with you? I hope you
don't think I mean to do you any harm."

"I am not afraid of you, but I choose to be
alone in the boat."

With this she pulled away from the shore, though
he continued to call out to her as long as she was
within hearing. She did not like the young man
at all, but rather despised than feared him. He
had often thrown himself in her way, and exerted
himself to please her. She was civil to him, and
that was all.

Lily remained in the boat, pulling about the
little bay for over an hour. Walk had stood upon
the beach for at least half an hour, waiting for her
return to the shore. Then he had retired, and
the fair maiden supposed he had gone back to the

institute. When she had taken all the air and
exercise she thought she needed, she rowed back
to the shore. Just as she had driven the bow of
the flatboat as far as she could on the sand, Walk
Billcord rushed out from the bushes, where he
had concealed himself, and prevented her from
getting out of the boat.

She had put the oars under the thwarts, and
arranged everything inside of the boat, which had
delayed her a few moments. But as soon as she
saw her tormentor running to the waterside, she
attempted to leap out of the boat.

"No, you don't, my pretty maiden!" exclaimed
Walk, as he seized her by the shoulders, and
crowded her back to her seat in the stern.

Under the impetus of the force applied to her
by the young man, Lily dropped into the seat, and
was obliged to grasp the gunwale of the boat to
avoid being thrown into the water. The fair face
of the young lady was flushed with anger, as well
it might have been, for she had not suspected that
her tormentor would resort to violence.

She was not inclined to submit quietly to the
will of Walk, for she immediately drew out one
of the oars from under the thwarts, and poised it
in the air, as though she intended to defeat the

intentions of the reckless young gentleman even by meeting force with force.

Walk Billcord stood for a moment holding on at the prow of the boat, as though he was undecided as to his next step. Doubtless he felt that he had already passed the bounds of propriety, and appeared to be considering whether it was prudent to proceed any further. A glance at the glowing and indignant face of Lily increased his interest in the adventure, and he was not willing to leave her in the moment of her heightened beauty.

Lily was the daughter of a poor dependent of his father : at least, he so regarded her, and thought he had some right to subject her to his own whim. He wanted to row her about the bay, and talk with her ; and this was the extent of his present wishes. It was only a "bit of a lark," a harmless pleasantry, on his part, as he afterward explained it, and he had not the slightest intention of injuring her.

The fair maiden did not regard herself as a proper subject for the young gentleman's pleasantry, and she was prepared to bring down the blade of the oar upon his head if the occasion should require. In the attitude of defence she waited for his next demonstration. The upraised oar rather tempted Walk to proceed, and he pushed the bow of the boat from the sand, springing into the foresheets as he did so.

As this was not a direct assault upon her, Lily did not bring down the oar upon his head, as she would under greater provocation, but she dropped it into the water at the stern of the boat. The water was shoal; and, setting the blade upon the sand at the bottom, she dexterously whirled the craft about, bringing the stern within a few feet of the dry sands on the shore.

Mr. Walk Billcord did not object to this movement, as it was necessary to head the boat away from the shore; but he deemed it prudent to secure the other oar before his fair companion could do so. He stooped down and got hold of the blade end of it. It required a little tact to remove it from its position under the thwarts; and, while he was engaged in doing it, Lily gave the oar another push, forcing the boat close up to the shore.

Without waiting for her tormentor to get the second oar over the forward thwart, she leaped lightly upon the dry sand, effecting her landing without wetting the soles of her shoes. She still held the oar in her hand, and stood on the shore, waiting for the next move of her unwelcome companion.

She was too proud to run away from such a contemptible being as she considered Mr. Walk Billcord. She looked as though she felt abundantly able to defend herself from any attack on the part of the unmanly persecutor. She evidently believed that he had no serious intention to harm her, but was simply making her the sport of his whim.

The moment she leaped ashore, Walk realized that she had got the better of him. Whatever he intended, he did not like to be outdone by a feeble girl. It was not pleasant for him, even in fun, to be outwitted by a weak maiden. He felt that he had not been smart, and he was annoyed at the situation. His vanity demanded that he should do something to get "even" with his intended victim.

The confident look and attitude of Lily on the shore disconcerted him, and invited further action on his part. He had not yet obtained possession of the oar, for it had to be shoved back before it could be passed over the forward thwart. But he had no present need of the implement, and he abandoned it to survey the position of Lily. He interpreted her looks and attitude as a defiance.

The boat, detached from the sand, was floating away from the shore. With a long leap he planted his feet on the land, and the effect of his movement was to drive the boat farther from the beach. A gentle breeze from the westward was driving it farther away, and Lily saw that it would soon be out of her reach.

She rushed to the water's edge, and, reaching out as far as she could, she succeeded in placing the end of the blade on the prow. She began to draw the truant craft toward the shore, when Walk put himself at her side. He took the oar from her hand, and pulled the craft up till its bottom grated on the sand.

Lily took a stick, and tried to get hold of the painter. As soon as she had it in her hand, Walk took it from her. He not only took the rope, but the hand which held it. He grasped her wrist with one hand, while he tried to drag the boat ashore with the other. He soon found that he had his hands full, both literally and figuratively.

Lily attempted to shake him off; but Walk tightened his hold upon her wrist, though he had to drop the painter of the boat, which, having no hold upon the land, began to float off into the open lake. The fair maiden turned and twisted in her efforts to escape, but the young ruffian held on like a vise.

In a moment or two she was exhausted with the violence of her exertions, and by this time she was thoroughly frightened. Very likely Walk had no worse intentions than at first, and was simply engaged in the business of getting "even" with the weak maiden who had outwitted him.

"What do you mean, you wretch? Let go of me!" gasped Lily, her chest heaving with terror and emotion.

"Don't make a fuss, my pretty one; I will not harm you," replied Walk.

"Let go of me, Mr. Billcord! I thought you claimed to be a gentleman! Let go, or I will scream," panted Lily.

"I only want to take a little row with you, and I shall, you may depend upon that," added Walk, picking up the oar which had fallen on the beach. "Don't make a fuss, and I won't hurt you."

But Lily again renewed the struggle with all her might. Just at that moment, Paul Bristol and his mother came out of the cottage. The boy was a stout youth of fifteen, and, the moment he saw what was going on, he broke into a run.

CHAPTER II.

PAUL BRISTOL seemed to have made only a couple of bounds before he had covered the distance between the cottage and the shore. He saw his sister struggling to release herself from the grasp of Walk Billcord. All the indignation Nature had portioned out to him was roused, and he did not stop to ask any questions. He did not even utter a word of warning or reproach.

His two fists were clinched in hard knots before he reached the scene of the encounter, and, without waiting to consider the situation, he planted a blow with his right fist between the two eyes of his sister's persecutor, and then did the same with the left. The effect was instant and decisive. Walk went over backwards upon the sand, and his hold upon the fair maiden was released.

By this time Mrs. Bristol had come to the spot, and, putting her arm around her panting, trembling daughter, she led her to the cottage without

25

taking note of the result of the battle, though she could not help seeing that the tormentor had been vanquished in the first onslaught.

Walk Billcord was utterly astonished as well as effectually upset. Paul Bristol had always been meek and subservient in his dealings with the students, and no one could have suspected that there was anything like a claw in his hard paws. If Mr. Walker was astonished the first moment after his unexpected fall, he was indignant and boiling over with wrath the second.

Though it was probable that both of the young gentleman's eyes had been put into mourning for the coming week, he was not otherwise damaged, and he leaped to his feet as soon as he could realize what had happened. He saw that he had been struck down by one whom he had always regarded as a son of toil, — a sort of cur about the premises of the institute. His blood boiled, and, without a word of any kind, he proceeded to "pitch into" his late assailant with all the physical vigor he could bring to bear upon him.

Paul warded off the wild blows aimed at him, and soon planted one of his own on the end of the young gentleman's nose, which caused the blood to flow in a stream from that organ. But Walk did not mind this little incident, though Paul was rather startled to see what he had done. The latter was inclined to deal as gently as he could with his gentlemanly opponent; but he found it necessary to defend himself from the impetuous charges of Walk. In doing so he delivered a hard hit, which carried his foe to the ground again.

The young gentleman was not yet satisfied, though he realized that he was not a match for his toil-hardened opponent. He sprang to his feet once more, out of breath, but unwilling to yield a hair to such an assailant. Grasping the stick Lily had used to haul in the boat, he again rushed upon Paul, and aimed a blow at his head; but Paul retreated a few steps, and picked up the oar which had dropped on the beach.

Paul Bristol was entirely cool, now that his sister was no longer in peril, and he began to realize that a quarrel with the son of the proprietor of the domain was a very serious matter. With the oar he warded off the blows of his insane adversary, and this was all he wished to do. He could easily have " laid him out " again, but the fear of consequences kept him within bounds.

Walk exhausted himself to no purpose. He could not hit his opponent, and his strength and his wind were soon used up. He drew back a little, and fixed a savage gaze upon his stalwart enemy. He panted like a wild beast at bay, and his blood boiled all the more because he could accomplish nothing.

"I'll settle you yet, Paul Bristol!" exclaimed Walk as he stepped down to the edge of the water and began to wash the blood from his face.

"I'm settled now," replied Paul calmly. "I have had enough of it, and I should like to stop where we are."

"You won't stop where we are, not till I have beaten you to a jelly. I shall break every bone in your dirty carcass before I get through with you," gasped Walk, struggling for an even supply of breath.

"When I say I have got enough of it, that ought to end the affair," added Paul with a cheerful smile on his face.

"I don't care what you say; you haven't got enough. You have given me two sore eyes and a bloody nose, and you haven't got anything to balance it," growled Walk. "I mean to break your head, and then I will call it square."

"But I don't want my head broken, if it is all the same to you," replied Paul, leaning on the oar. "My head is of some use to me, and it would not be pleasant to have it broken."

"You began it, and you shall have enough of it before we are done," added Walk, beginning to breathe a little more freely.

"I began it?" queried Paul with the same cheerful smile. "I don't think so, and I should like to argue the question with you."

"Didn't you hit me first, you nunkhead?" demanded Walk.

"Didn't you lay hold of my sister first, and frighten her half out of her wits?"

"I didn't hurt her, and I was only fooling with her."

"Fooling with her! That's just what I was doing with you. I was only fooling with you, Mr. Walker."

"I don't like that sort of fooling, you speckled cur!"

"My sister didn't like your sort of fooling any better than you like mine. But, if you want to stop fooling, now is a good time to begin."

"I will stop when I get even with you, and not before," snapped Walk. "You struck the first blow, and I mean to strike the last."

By this time the young gentleman had fairly

recovered his wind, but nothing like coolness had come over his temper. Dropping the stick, he rushed upon Paul again with his naked fists. He was savage, and the boatman's son soon found that he could not passively defend himself, and the result was that Walk soon went under again.

This disaster made him madder than ever, and when he rose from the beach he seized the stick again, which Paul met with the oar. Paul liked this way of carrying on the combat better than the other, for he could defend himself without inflicting any injury on his furious opponent.

While Walk was thus wearing himself out, a gentleman with a riding-whip in his hand came out of the path through the woods. As soon as he discovered what was going on upon the beach, he quickened his pace, and reached the scene of the conflict at a sharp run. It was Major Billcord, the father of Paul's wrathy opponent.

"What does all this mean?" demanded the major when he had come within speaking distance of the combatants. "How dare you strike my son with that oar?"

"I haven't struck him once with it," replied Paul, aghast at the presence of the mighty proprietor of the domain. "I am only defending myself, sir."

"You have no business to defend yourself against my son, you dirty puppy. How dare you lift a weapon against him?" stormed Major Billcord; and to him there was only one side to the controversy, whatever it was.

Walk had dropped his stick as soon as he heard the voice of his father, and Paul had done the same with the oar. The latter felt that he had got into a very bad scrape. The major was a magnate of the first order, and he was supreme on his own domain. His mother was a tenant at will at the cottage. All the money she had inherited from her father's estate, and all she had in the world, was invested in that cottage. The mighty major could turn them out of house and home at a moment's notice, as they paid no rent.

"What does all this mean, my son? I am sorry to see you fighting with such a cur as that," said Major Billcord when the battle was suspended for the moment.

"It means that he struck me first, and I intend to get even with him if I fight till Lake Champlain dries up," blustered Walk, as he clinched his fists again; and doubtless he had a clear idea of his father's views on the subject of pugilism.

"He struck first! You did quite right, my son.

Never take a blow from any one," added the major.

"But he insulted my sister, sir! He had seized hold of her, and held her when I hit him, sir," pleaded Paul with proper deference; and he felt that he had a good defence.

"A fight begins with the first blow, and we need n't ask what happened before it was struck. You admit that you struck the first blow, Bristol?" continued Major Billcord, sitting in judgment on the case.

"I did strike the first blow, sir; and a fellow that would n't hit hard when his sister was insulted, and held as a prisoner, don't amount to much," Paul replied rather warmly.

"You struck the first blow; and that 's all I want to hear about it," added the major sharply. "My son has done quite right to resent a blow with another blow; and if he is not satisfied with the punishment he has given you, you vagabond, I will stand by and see fair play till he is satisfied."

Mr. Walker did not quite approve the ground taken by his father, and wanted him to do something more than stand by and see fair play. But the major had spoken, and the son realized that

he had nothing to do but to take the broad hint
the patriarch had given him. Clinching his fists
again, he rushed upon Paul for the third time.
Paul was indignant at the decision of the magnate,
and felt as though he had been commanded by the
great man to permit his son to insult his sister.

Walk rushed upon him, but Paul's back was up
for the first time since he had relieved his sister
from the grasp of her assailant. His paws were
not velvet: they were all fangs. At the first onslaught of Walk, that young gentleman went over
on his back with the blood gushing from his nos-
trils. Twice more he renewed the attack, with
about the same result.

Mr. Walker was so full of wrath that he could
no longer control himself, and he laid hold of the
stick again. Paul picked up the oar once more.
The son of toil knocked the stick out of the hands
of his opponent, and it flew into the lake. Walk
could not find another, and Paul dropped the oar.
It was naked fists again, with the same effect as
before.

By this time Major Billcord was as full of wrath
as his son, and without regard to fair play, of
which he stood as champion, he rushed to the assistance of his defeated son. Paul picked up the
oar and retreated before the two.

"Stop a moment, if you please, Major Billcord," shouted Paul. "I don't want to hit you sir, and I won't if I can help it."

"But I am going to flog you within an inch of your life!" yelled the major.

Paul had gone as far as he could without retreating into the cottage, and he was unwilling to carry the battle into the presence of his mother and sister. He halted; the major wrenched the oar from his grasp. He struck the son of toil with it. Paul's blood was up; he gave the magnate a blow between the eyes, under which he went down. Walk "pitched in" again, and was planted by the side of his father.

CHAPTER III.

MAJOR BILLCORD was a short, puffy man, inclined to corpulency. The blow of the son of toil, and his fall upon the sand, proved to be enough for him. He was all foam and fury in consequence of his signal defeat. Possibly he had thought that a poor dependent upon his bounty would not dare to strike him; and, in truth, Paul felt that it was something like treading upon the Bible.

He had attempted to take the stalwart youth by the collar, and had struck him with his riding-whip in a tender place. The pain was nothing, but the indignity was great; and Paul's impulse had led him farther than he would have gone if he had considered what he was doing.

The major and his son picked themselves up, and for a moment they gazed with something like wonder upon the victor in the unequal contest. But all three of them had been beside themselves

35

for the moment.   Paul realized what he had done ;
and so did his mother and sister, for they came
out of the cottage while father and son were getting up from the ground.

"Woman, do you see what your son has done?"
demanded Major Billcord, who was the first to
break the impressive silence.

"I am very sorry, sir," pleaded the poor woman,
stepping between Paul and his victims, in order to
prevent him from doing them any further mischief
if he should be disposed to renew the combat.

"Sorry for it!" exclaimed the magnate, as if
simple regret could atone for a blow given by a
plebeian to a patrician. "Is this the way you
bring up your son?"

"I am very sorry, Major Billcord, but he has
been greatly provoked. By your leave, sir, it
was Mr. Walker that began it."

"It is false, marm! Your brute of a son struck
the first blow ; he has confessed it to me," puffed
the magnate.

"But Mr. Walker had first insulted my daughter ; he had seized hold of her, and was trying
to force her into the boat when Paul interfered,"
Mrs. Bristol explained with as much meekness as
the subject would permit.

"Nonsense, woman! Seized hold of your daughter! Don't talk such stuff to me. Walker did not mean to do her any harm," added Major Billcord with the utmost contempt.

"I only asked her to let me row her about the bay in the boat," the young gentleman explained.

"It was impertinent in her to refuse when my son honored her with his notice," continued the major.

"I thought she had a right to choose her own company," said Mrs. Bristol with proper humility.

"I have allowed you to live on my land for two years without a penny of rent, woman; and this is the return I get for it," replied the great man, in whose heart the poor woman's ingratitude was beginning to make havoc.

"You have been very kind to us, Major Billcord, and we are very grateful for all you have done for us. I am so sorry that this sad thing has happened!" pleaded Mrs. Bristol.

"And still you try to fasten the blame on my son," retorted the proprietor of Sandy Point and its surroundings.

"I am very sorry he meddled with Lily; if he had n't done it, there would have been no trouble, for Paul has always treated Mr. Walker with respect."

"At it again !" exclaimed the major. "You will insist that my son was to blame, simply because he was polite enough to invite your daughter to take a row with him in the boat."

"She was not willing to go ; and I did n't know that she was obliged to go out on the lake with him. She declined his invitation, and Mr. Walker tried to force her into the boat."

"It was not civil in her to decline the invitation, and I don't wonder that Walker was a little vexed at her refusal. She is a pert minx, marm, and has not been well brought up, or she would have known better than to decline," added the magnate, bestowing a look of severity upon the fair maiden.

Mrs. Bristol and Paul saw that it was useless to attempt to reason with such a man, and they were silent. The major took out his handkerchief, and wiped the perspiration from his face. Then he felt of his nose and the region about his two eyes, between which the son of toil had planted his hard fist. Doubtless there was a soreness in those parts, and perhaps the visual organs of the father would be clothed in sable wreaths by the next day.

"That boy must be punished, severely punished,

for what he has done," the major resumed. "He has had the audacity to strike me in the face,—me, the benefactor of the whole family!"

"Did n't you catch me by the throat, and hit me with your riding-whip, sir?" asked Paul calmly and meekly.

"What if I did! Do you mean to put yourself on a level with me, you young reprobate?" demanded the magnate, his wrath beginning to boil again. "Woman, I say that boy must be severely punished for this," he continued, turning to Mrs. Bristol again. "He must be whipped till he can't stand up!"

"Who will whip him, sir?" asked the poor woman innocently.

"I will do it, if you don't, marm," replied the major savagely.

"I could not whip him, sir; he is a great deal stronger than I am; and, if he is whipped at all, you must do it, sir;" but Mrs. Bristol seemed to think there was something a little satirical in what she said.

"Then I will do it!" said the magnate, raising his riding-whip.

"Perhaps he will not allow you to whip him, sir," suggested Mrs. Bristol; and even her anger appeared to be approaching the boiling-point.

"The boy deserves to be severely punished. If he submits to the whipping which Walker and I will give him, we may be willing to let the matter drop where it is."

"You had better arrange it with Paul, sir. I should as soon think of whipping Colonel Buckmill as my son," replied the poor woman with a decided touch of satire in her tones and manner.

"If the young villain submits, very well."

"If you should begin to punish him, I have no doubt he will speak or act for himself," she added.

"Bristol, you hear what has been said. Will you submit to the punishment you deserve?" demanded the major severely, turning to the culprit.

"No, sir, I will not."

"Do you hear him, marm?"

"I do, sir; and he answers just as I supposed he would."

"Then you uphold him in his treacherous treatment of my son? Then you countenance him in biting the hand that feeds him?"

Mrs. Bristol made no reply, for she did not wish to irritate the powerful man unnecessarily. She looked at her son, and she was proud of him.

"Bristol, you refuse to submit to the whipping you deserve?" demanded Major Billcord, approach-

ing the stout youth with the riding-whip upraised.

"If you hit me with that whip, sir, I will knock
you as far beyond the middle of next week as I
can," replied Paul firmly and quietly. "Your son
insulted my sister, and I treated him as he deserved, and just as I would another time if he did
the same thing. My sister is a poor girl, but she
is just as good as you are, and just as good as Mr.
Walker is. If she is insulted, sir, I will stand up
against five hundred Billcords as long as there is
anything left of me."

"Is this your gratitude for what I have done for
the family?" asked the major, knitting his brow
into a knot of wrinkles.

"Yes, sir; this is my gratitude. Do you think,
because you allowed my father to put his cottage
on your land, that you and your son have the
right to insult my sister?" demanded Paul with
considerable energy.

"No one insulted her, you young reprobate!"
interposed the father. "Is a civil and gentlemanly
invitation an insult?"

"If he had stopped there, we should have had no
trouble."

"But she refused the invitation."

"She had as much right to decline it as any lady in Westport would have."

"Was it treating a member of my family properly, after all I have done for you?" demanded the major more calmly, but with a terrible havoc in his tender feelings.

"You have had a good deal to say about what you have done for us, Major Billcord. The land on which that cottage stands," continued Paul, pointing to it, "is not worth ten dollars. At ten per cent, the ground rent would be one dollar a year, or two dollars for the two years it has stood there. I have done work enough for you in the shape of errands, taking care of your boat, and in other ways, to pay for the land twice over. I have carried the first black bass of the season to your house, when I could have sold the fish for a dollar apiece, for two years. As I look at the question of gratitude, there is a balance of at least twenty dollars in my favor; but I give it to you with all my heart, and I don't claim the privilege of insulting your daughter for what I have done."

"You are a glib-talking puppy, and there is no more reason or common sense in you than there is in a heifer calf. I have had enough of you, and so has my son," responded the major, choking with

wrath over the unanswerable argument of the poor
dependent.

"If you have had enough of me, you and Mr.
Walker, I am satisfied to let the matter drop where
it is; but if Mr. Walker, or any other student of
the Chesterfield Collegiate Institute, insults my
sister, I shall hit him as hard as I can," replied
Paul coolly.

" Woman, you have heard the insulting words of
your son, and you uphold him in his wickedness.
I must take the next step. I will not have such a
vile reprobate on my land. I will not have you or
your ungrateful daughter on my territory. You
are a tenant at will. That cottage must not remain another day on my premises. Remove it at
once. If it is here at three o'clock to-morrow afternoon, I will give the students permission to tumble
it into the lake. Do you hear me, woman?"
stormed the major fiercely.

"I hear you, sir," replied Mrs. Bristol, covering
her face with her handkerchief, and beginning to
weep bitterly.

" You needn't cry about it, marm. You and that
graceless son of yours have brought it on your-
selves; and I think the students will enjoy the fun
of pitching the shanty into the lake."

"It is all the property I have in the world, Major Billcord," pleaded the poor woman. "Give me a little time to remove the cottage, I implore you!"

"Don't implore me, marm! Thank your wretch of a son for it. By three o'clock to-morrow afternoon, if you have n't removed it in the mean time, the shanty shall be rolled into the lake."

"I cannot get it through the woods to remove it," groaned Mrs. Bristol.

"That's your lookout, marm," said the major as he and Walk departed.

Mrs. Bristol seated herself on the lower step of the cottage, and continued to weep bitterly.

CHAPTER IV.

THE cottage of the Bristols had been framed in Burlington, and brought down to Sandy Point on a schooner. As it stood, it was estimated to be worth about three hundred and fifty dollars, which was the cost of it to the poor woman when she invested her all in what was to be a home for the family.

It was a small sum the cottage cost, but to the poor woman it was as big as a million to a millionnaire. She had been well brought up in her father's house, and she could not exist like a Chinaman or a Hottentot, and it had cost the family a struggle to live during the absence of the father.

Now all that she had was to be taken from her. As they had paid no ground rent for the site, the law could do nothing for her. She was a tenant on suffrance rather than a tenant at will, and had no rights whatever. The magnate could tumble the cottage into the lake, and the wind would carry

45

it where it listed. It would probably be broken up on the rocks or shoals, and the major might as well set it on fire as turn it adrift on the lake.

The rich man intended to execute his mandate in the cruelest manner possible. The students were to have a frolic in tumbling it into the lake. The humble structure contained all their household goods, all the little articles they valued far beyond the money they cost. It was hardly possible to remove them in the time allowed for the purpose, for everything would have to be carried by hand or transported in the flatboat.

No team could be driven down to the point, for the major would not allow a tree to be felled to make a road, and the owner had been compelled to leave his saddle-horse at a considerable distance from the lake when he visited it. Of course, the cruel magnate understood all this, and realized that his final mandate doomed the cottage and all it contained to certain destruction, for neither he nor his persecuted tenants could see any means of relief.

Even if they could carry away their goods, they had no place to put them. The brief period of probation given them was not more than enough to enable the poor woman to find another tenement.

It was two miles to Westport, and five to Genverres, by water. The situation looked entirely hopeless to Mrs. Bristol; and the more she thought of it, the more bitterly she wept.

"I don't know what will become of us," said she when she had vented her grief for a time.

"Don't cry, mother: we shall get out of the scrape in some way," replied Paul in as soothing tones as he could command, for the situation was hardly more hopeful to him than to his mother.

"I don't see that we can do anything but submit to the loss of everything we have," moaned Mrs. Bristol. "We can't stay here any longer, and we have no place to go to in the wide world. The students will take a wicked delight in breaking up everything we have. I cannot stay here to see them revel in the destruction of our home, which has been as dear to me as though it had been a palace. But where can I go?"

"We had better go to Westport, mother," said Lily, wiping the tears from her eyes. "We can take a few things with us in the boat."

"The boat went adrift in the row, and I saw it halfway over to Scotch Bonnet," interposed Paul as he strained his eyes to discover the truant craft.

"Then we can walk over to Westport; but we

can't carry much of anything in our hands in a walk of two miles," added Lily gloomily.

"Where shall we go in Westport when we get there?" asked Mrs. Bristol. "We can't go to a hotel or boarding-house, for we have n't money enough to pay our way for three days."

"I don't see that we can do a thing, mother," said Paul when he had carefully looked over the situation. "I would borrow a boat, if there was one to be had; but I am sure the institute boatman would not lend me one now. Major Billcord's story will be all over the neighborhood in a few hours. I could get one in Westport; but it might take me a whole day to find our flatboat, for it must have been driven ashore on the other side of the lake. Some vessel may have picked it up, for I saw two or three going up the lake."

"I saw a sailboat go by while you were talking to the major," said Lily.

"I noticed her; it was one of the Beech Hill boats," added Paul. "One of these vessels may have picked up the 'Dragon,'" as he had named the flatboat, "and it may be five miles from here by this time."

"We are in the hands of Providence, and as helpless as babies," continued Mrs. Bristol. "I don't see that we can do a single thing for ourselves, and we must trust in a higher power than man."

"We can stay in the cottage over night, at least, and it will be our last in our happy home," said Paul. "The students will all be at their studies in the forenoon, and then you and Lily can walk over to Westport."

"What are you going to do, Paul?" inquired his mother, bestowing a look of the deepest interest upon him.

"I shall stay here and save what I can."

"You must not stay here!" protested his mother warmly. "The students will kill you, under the lead of Walker Billcord."

"I am not afraid of them."

"You must not stay here: if you do, I shall stay with you," replied the devoted mother.

"There is that sailboat, Paul," said Lily, who had been gazing blankly out upon the lake. "She looks as though she was headed for the point."

"That's the 'Goldwing,'" added Paul. "I hope Dory Dornwood is on board of her. He is a good fellow, and he may do something to help us. If she comes near enough, I will hail her."

"But Dory Dornwood will not do anything for

any one on this side of the lake," replied Mrs.
Bristol. "The two schools have waged the fier-
cest war upon each other."

"I know Dory Dornwood very well, mother. I
had a long talk with him about the war between
the schools, and I know that the Beech-Hillers
have done all they could to keep out of trouble
with the Chesterfields. I am sure he will do anything he can for us. He don't like Major Billcord
any better than I do at this moment, for he had a
row with him when he was a waiter on a steamer."

"I don't see that he can do anything for us, even
if he does come near enough for you to hail him,"
added Mrs. Bristol, hardly less despondent than
before.

"He can take some of our things on board his
boat, and carry them to a place where they will be
safe until we want them again."

Paul was quite hopeful that something would
come of the visit of the "Goldwing" to the point,
if she came there. Without stopping to argue
the possibilities with his mother, he hastened to
the shore. The "Goldwing" was headed down the
lake, and, with all sail set, she was dashing over
the waves at a tremendous high speed. She was
towing a small boat astern of her, but Paul could

not tell whether it was the "Dragon" or her usual tender.

The wind was exactly west, and the schooner was going free. When Paul saw her before, she was on the easterly side of the lake, where she got a better breeze than on the opposite side. He was satisfied that she headed for the vicinity of the point; for she would have started her sheets when she came up with Scotch Bonnet, if she had been bound directly down the lake.

Though she was approaching the point, it was by no means certain that she intended to make a landing there. The boat she was towing was astern of her, and Paul could not see it plainly, as the hull of the yacht obstructed his view. As on all occasions, he determined to do the best he could. Running back to the house, he borrowed a tablecloth of reddish hue, and fastened it to the oar. Elevating it as a signal to the "Goldwing," he held it in position at the very apex of the point.

Lily and her mother were not inclined to join him, for they did not feel in the mood to meet any young men, however civil they might be. They remained seated on the steps of the cottage; but they watched the sails of the yacht with as deep an interest as Paul did, for it was at least possible that the party on board of her might help them in their present dire emergency.

"She is headed for the point!" shouted Paul, when he had satisfied himself of the fact.

"I pray that a kind Providence has sent her to our relief," replied Mrs. Bristol.

Paul watched her with the most intense interest. When the "Goldwing" was within a quarter of a mile of the point, the party on board of her waved their handkerchiefs as a reply to Paul's signal. The hopes of the watcher on the shore ran high, but he could not yet see whether or not she had the "Dragon" in tow.

Sandy Point was a kind of double cape. It was shaped like a letter T. The cottage was on the northeast point, inside of which was Sandy Bay, where the Beech Hill students sometimes came to bathe. The other arm of the point extended to the southwest, and inside of it was the station of the boats belonging to the institute.

"Is that you, Paul Bristol?" shouted some one on board of the "Goldwing."

"Yes! Is Dory Dornwood on board?" replied Paul.

"He is! Have you lost a flatboat?" called the speaker from the yacht.

"I have!"

The schooner kept well off the point, and appeared now to be headed for the shore on the opposite side of Sandy Point. As she came abreast of the shore, Paul saw that she had the " Dragon " and her tender in tow. The return of the " Dragon " was a godsend, even if nothing else resulted from the visit of the yacht. The " Goldwing" ran over towards the opposite shore, and then tacked. The wind was light inside of the bay, and the schooner circled gracefully about, coming up into the wind off the point where Paul stood. Down went her jib in the twinkling of an eye, and over went her anchor. In a moment she had come up to her cable, with her mainsail fluttering in the breeze.

One of the party hauled up the " Dragon," and, casting off the painter, sculled her ashore with the single oar that remained on board of her.

"This is your boat, I suppose," said Tucker Prince, one of the new students of the Beech Hill Industrial School, as he stepped on shore with the painter in his hand.

"It is my boat, and I owe you a thousand thanks for bringing her back to me," replied Paul.

" The skipper wants to know if any accident has

happened by which she was turned adrift," continued Tuck Prince, whom the New-Yorkers called
a " Bosting boy."

" No accident; but who is the skipper?"

" Dory Dornwood."

"I would give more to see him than I would to
meet my grandmother," added Paul.

In a few moments more Paul Bristol had Dory
by the hand.

CHAPTER V.

DORY DORNWOOD had been to the point before, and had met Paul Bristol there. Of course, the son of toil had never associated with the Chesterfields on anything like an equality, and he had no especial interest in them or their affairs. In the quarrels between the two schools he had been entirely neutral, for the reason that he had not been called upon to take part in any of the battles, and his opinion of the merits of either side was of no consequence..

Dory gave the resident of the point his hand, as he came on board of the "Goldwing," and was glad to see him. Possibly the skipper was willing to have a friend in this particular locality, though he knew that Paul had no influence with the gentlemanly students of the institute.

"I picked up your boat over by Scotch Bonnet," said Dory. "It was just going on the rocks, and I thought it would have a hard time there. I

55

knew from the direction of the wind that it must have come from this region, and I was afraid some one had been upset in her, for there was only one oar in the boat."

"No one was upset in her, though the 'Dragon' is not inclined to stay right side up when the sea is heavy," replied Paul. "I am more thankful to you than I can tell for bringing her back at just this time. If you are willing to hear me, I should like to tell you how she happened to get adrift, and why I am so glad to get her back."

"I am in no hurry, and I am ready to hear any-thing you wish to say to me," replied Dory, as he invited Paul to take a seat in the standing-room.

There were five other Beech Hill students in the boat, and they seemed to be as willing to hear the story as the skipper was. All of them judged by the manner of the Sandy-Pointer that he had something of interest and importance to tell. Paul gave his account of the trouble between Lily and Walk Billcord without exaggeration or embellish-ment, though he did not do justice, in his modesty, to his own prowess in the battles with the magnate and his son.

The narrative was exciting enough to secure the

closest attention of his auditors; and, when it was finished, all of them had some questions to ask, and most of them some comments to make. As they were not prejudiced in favor of the major or his son, they agreed that the son of toil had served them right. Dory said he should have been very sorry to strike a man of the age of Major Bill-cord, but he did not see how he could have helped doing so under the circumstances.

"And he has ordered you to remove the cottage at less than a day's notice?" continued the skip-per.

"He has, and at a time when my boat had gone adrift, and I had no way to move a single thing which we could not carry two miles in our hands. That is the reason why I was so glad to see the 'Dragon' coming back to the point," replied Paul, looking at the skipper of the "Goldwing" with the deepest interest. "But the water is rather rough this afternoon out on the lake, and I am afraid I can't do much towards moving my mother and sister and all the furniture in that flatboat."

"Then you mean to leave the cottage to be pitched into the lake?" added Dory.

"I don't see that we can do anything else," replied Paul blankly. "It is about a quarter of a mile through the woods to the road, and Major Billcord will not allow a tree to be cut down. I could not move it if I had a hundred men to help me. I might take it to pieces, if I had time enough to do the job; but we have little time, and not much money."

"Suppose we go on shore and take a look at the cottage," suggested Dory, who seemed to be doing some heavy thinking, though he did not hint that he knew of any remedy for the misfortunes of the Bristols. The tender was brought up to the gangway, and Dory went ashore in it, accompanied by Tuck Prince, while Paul took the "Dragon" back.

The skipper walked up to the cottage, and Paul introduced him to his mother and sister. Dory had never seen Lily Bristol before, but he was perfectly willing to agree with the "speech of people," that she was the prettiest girl in the State of New York, and possibly in the State of Vermont. But she looked very sad, and so did her mother, and Dory said nothing to them about the mandate of the magnate.

The visitor looked at the house, and got its approximate dimensions in his mind. He examined the foundations of the structure, and the land upon which it lay. He was thinking, but he said nothing.

"Of course, Major Billcord knows very well that we cannot move the cottage without making a pathway for it through the grove," said Paul, who had observed the expression on Dory's face with the utmost interest.

But he had not expected that the skipper of the "Goldwing" could do anything about moving the dwelling : the most he had hoped was that the visitor would volunteer to assist in the transportation of the furniture, or a portion of it, to some safe place, if such could be found. But Dory was provokingly silent, and did not hint at anything. When he had completed his examination of the premises, he walked towards the end of the point again.

"I suppose you will agree with me that not a thing can be done," said Paul as he followed the skipper; and he began to be very much discouraged at the prospect.

"I don't know whether anything can be done or not," replied Dory, musing. "I can't do anything myself, for I don't feel at liberty to act without the knowledge of my uncle, Captain Gildrock," replied Dory. "If you could move the cottage, have you any other place to put it?"

"Not a place. My mother owns a house, but

not a foot of land in the wide universe," answered Paul gloomily.

"It is not very easy to think of any plan for moving the building when we have n't any idea of what is to be done with it," added Dory, surveying with his eye the ground between the cottage and the water of the lake.

"I did n't suppose you could do anything for us in that way," continued Paul. "I thought we might save some of our furniture in the 'Dragon.'"

"You could n't even get a bureau on board of her," replied Dory, glancing at the flatboat. "If you laid it across the gunwales, it would upset her. But I have an idea, though it may not amount to much. What are your mother and sister going to do? I take it they will not stay here to see the Chesterfields destroy your cottage and its contents?"

"No; they will leave in the morning; but there is no place under the sun for them to go. We have no relations, and hardly any friends," answered Paul very sadly.

"As I said before, I can't do anything without orders in a case like this. I am very sorry for you. If the situation were what my uncle defines as an emergency, I should be permitted to do what is required to save life or property. But there is nearly a whole day's leeway in this case," continued the skipper.

"I didn't know but you might carry some of our things over to the other side in the 'Goldwing,'" Paul hinted very timidly.

"I might be able to take some of them; but the schooner would not hold one-half of the goods. I hope to do something better than that, though I can't promise anything. So far as the furniture is concerned, I don't think you need to worry about that, for there will be time enough to remove it to-morrow morning," said Dory.

"Plenty of time, if I have any help," added Paul.

Dory hoped he should be able to do something better than this; and the words had given Paul a strong hope, though he wished the skipper would speak out plainly, and say what he thought of doing.

"It is now about five o'clock," continued Dory, looking at his watch. "We have plenty of time, and I have something to propose. I can't do anything without my uncle's knowledge, but I should like to have your mother and sister go over to Beech Hill in the 'Goldwing;' and you too, Paul, if you don't think it necessary to remain here and look out for the cottage."

"What should they go over there for?" asked the son of toil.

"To tell my uncle your story. If Captain Gildrock decides that nothing can be done for you, I will bring your mother and sister back before dark. Then I will take a cargo of your goods to any place you say in this part of the lake. That is all I feel at liberty to do under the present circumstances," added Dory. "But I am pretty sure that something more will be done for you."

"I will speak to my mother about it," replied Paul. "Will you come with me, Dory?"

The skipper followed him to the cottage, and Paul stated the case to his mother. She was willing to do anything her son thought best. She did not think it was necessary for Lily to go if Paul was to remain in charge of the house; but Dory thought she was a part of the story, and anticipated some hard questions from his uncle which she could answer better than any other person. Mrs. Bristol yielded the point, and in a few minutes they were ready for the trip.

The ladies were seated in the standing-room, and the Beech-Hillers were as polite as so many dan-

cing-masters, "tinkers" though they were. A
short distance from the shore the wind was still
fresh; and in half an hour the "Goldwing" was in
Beechwater, as the principal of the school generally
called the lake.

Without waiting to moor the schooner, which he
left in charge of Tuck Prince, Dory pulled the
ladies to the new boat-house in the tender. They
landed at the steps, and the skipper conducted them
to Captain Gildrock's library. They were pleas-
antly welcomed by the principal, though they were
entire strangers to him.

Dory stated that he had brought Mrs. Bristol and
Miss Lily from Sandy Point, and he wished his
uncle to hear the story they had to tell. He hinted
that the visitors had better confine themselves to
the facts in the case, without any comments; and,
as Lily had been the principal and first actor in the
drama, he thought she had better open the narrative.

Possibly Dory thought an account of the opening proceedings from the lips of so pretty a girl as
Lily might have more influence with his bachelor
uncle. The captain smiled graciously, and bowed
encouragingly to the fair maiden. She began in a
very straightforward way with the narrative, and

Dory was glad the occasion permitted him to gaze at her without staring.

When she had completed her narrative, there was but little more for her mother to say. The principal asked a few questions, and then he was in possession of all the facts. He knew all about Major Billcord, and he had no difficulty in believing the simple and unadorned statement to which he had listened. Very likely he was as indignant as any Christian man would have been at the outrage of the magnate and his son, but he did not express himself in this direction.

When Mrs. Bristol and Lily had said all they had to say, the captain looked at Dory to see if he had anything to offer. Dory was not slow to take a hint, and he made quite an energetic speech of considerable length, setting forth his views of the situation.

CHAPTER VI.

A CALL FOR ALL HANDS AT BEECH HILL.

I AM very sorry, Mrs. Bristol, that you should have been placed in such an unpleasant situation," said Captain Gildrock when Dory had brought his speech to a square conclusion, which some orators find it very difficult to do. "Your son did no more than I would have done in the same circumstances. It was highly proper for him to defend his sister with his fists; and after that he acted only on the defensive."

"I was very well satisfied with Paul, sir," added Mrs. Bristol.

"I think you have reason to be. You have heard what my nephew has said, and I fully indorse the plan he has outlined. I shall leave it to him to carry it out in his own way."

"You are very kind, Captain Gildrock, and I shall be grateful to you as long as I live," replied Mrs. Bristol, with enthusiasm. "Paul says your students were always much better behaved than those of the institute."

"Unfortunately our relations with the school on the other side of the lake are not as pleasant as I could wish; but I do not intend that our young men shall be offensive to their neighbors."

"The Goldwing is all ready to take you back to Sandy Point, Mrs. Bristol," Dory interposed.

"But why should you return, madam?" said the principal. "I think you had better remain here. We have plenty of spare rooms, and we will do all we can to make you comfortable."

"Thank you, sir; but I am afraid Paul will be uneasy in my longer absence."

"Dory shall run over to Sandy Point, and inform your son what is to be done, and can bring you anything you may want," suggested Captain Gildrock.

The principal gave some strong reasons why she and Lily had better remain at the mansion over night, and she finally consented to do so. Mrs. Dornwood and Marian, Dory's mother and sister, were called, and they soon made the visitors feel quite at home. Dory returned to the Goldwing, and was soon standing out of Beechwater.

In less than half an hour the schooner was at the point. Paul was greatly astonished, and not a little troubled, when he saw that his mother and

sister were not on board of her. But the skipper
soon explained their absence, and stated what was
to be done with the permission of his uncle. Paul
went to the cottage for a few articles which his
mother had desired, in a note, while the skipper
looked over the situation of the cottage again, and
arranged his plans for action.

"We shall disappoint the Chesterfields once
more, Paul," said Dory, when the son of toil
joined him. "We have spoiled some of their little
arrangements before."

"They will miss the fun the major has promised
them, but I think he will feel the loss of it more
than they will. Of course, all he wants is to
punish us," replied Paul, with a cheerful smile.

"You can go over to Beech Hill with me if you
like," continued Dory, when they reached the
tender.

"I must stay here and watch the cottage. Some
of the students might think it was fun to set it on
fire to-night, though it would not make a very
brilliant light in the moonshine," replied Paul.

"Do you expect any of the Chesterfields at the
point to-night, Paul?" asked Dory with some
anxiety, for their presence might interfere with his
plan.

"No; I hardly expect any of them. I don't know that Major Billcord has told them about the fun in store for them yet, though he was so mad when he left the point that he could hardly keep it in," answered Paul.

" Well, if they come to-night, we can't help it," added Dory, as he stepped into the tender. " We shall be here all the same, and we shall do the work we have laid out."

The Goldwing got up her anchor, and filled away. Paul watched her till she disappeared in Beaver River. The situation had changed entirely, and Paul was as happy as though there had been no tempest at the point that day. His mother and sister were in good quarters, and he did not much care if the Chesterfields came down upon him in full force. As soon as the schooner was out of sight he went into the house to get his supper.

As soon as the Goldwing was moored, and her crew had eaten their supper, there was a call for all hands to assemble at the new boat-house. The famous structure had been finished nearly a year before, for it was in the month of May that the trouble at Sandy Point took place. The school year began in the September preceding.

After considering the subject during the summer, Captain Gildrock had decided to increase the number of pupils in the Beech Hill Industrial School. But he was a prudent and practical man, and he had taken only a dozen additional scholars. Two had left to take good-paying situations, and the whole number now was thirty-six. There was room in the enlarged dormitory for a dozen more, and space enough for them at the benches in the shops.

A third class had been formed of the beginners; and, as they had been under instruction for eight months, some of them had acquired considerable skill in the use of tools. Another barge had been procured, and the "green hands" had all learned to row, to swim, and a few of them to handle a sailboat. The school was now larger than the one on the other side of the lake. But the Chesterfields, after having been defeated several times in their assaults upon the Beech Hillers, had confined their attention more to their own affairs than formerly. They were satisfied to give the barges of the "tinkers" a wide berth on the lake; and the boating season closed without any more serious quarrels on the water.

The Topovers had never accomplished anything by meddling with the students on their side of the

lake. A little discipline in one of the courts had
kept them at a distance for a time. When the
fruit was ripe, Mr. Brookbine's big dog became a
terror to them; for the master carpenter had built
a house for him near the rear fence of the orchard,
and the animal understood his duty perfectly.

The call for the students after supper was unusual, and no one but the members of the acting
crew of the yacht knew what it meant; and even
they knew nothing of the plan they were to assist
in carrying out. Since the former season there
had been some changes in the organization of
the students. Captain Gildrock was no longer
the actual captain of the Sylph, the beautiful
steam yacht connected with the institution.

The position had been given to Dory Dornwood,
and the students generally sailed her without the
interference of the principal or any of the instructors. Mr. Jepson, the master-machinist, was
no longer the chief engineer, and was therefore at
no time under the orders of any of the juvenile
officers. Corny Minkfield, who had served one
season as first assistant-engineer, had been promoted to the highest place, and the second to the
place thus made vacant.

Oscar Chester was the first pilot. He had been

a diligent student in the pilot-house, and knew the lake almost as well as the captain. All the places had been filled after the first appointments in accordance with the merits of the students, though of necessity "civil service" rules prevailed, for the reason that the members of the ship's company had become more skilful in the departments in which they had been employed than in any other.

The only violent changes made were those which gave the cooks and stewards a chance to learn seamanship or the management of the engine and furnaces. As waiters they learned out in a few months, and even the rather limited routine of cookery required on board was exhausted in the same time. Old deck hands and firemen became stewards, while those who had served in the fire-room and cabins were transferred to the deck.

The increase in the number of students allowed a very large force of seamen, and the vessel was now heavily manned. Crews for the quarter boats were appointed for permanent service, and four quartermasters were added to the organization, who had regular tricks at the wheel in the pilot-house under the direction of the first or second pilot.

Dory Dornwood had been in command of the steamer for the three months at the close of the last season of navigation, and every Saturday he exercised his ship's company in as long cruises as the length of Lake Champlain would permit. Sometimes the principal was on board, and sometimes he was not. If he had anything to say, he said it to Captain Dory Dornwood; and the discipline was as perfect as though the steamer had been in the navy.

In the beginning of Captain Dornwood's administration there had been considerable difficulty. Boys from the country, or even from the city, were not very prompt to see the necessity of obeying orders without asking any questions. But as this was one of the principal lessons the steam yacht was to impart to the pupils, there was no relaxation of the discipline to accommodate those who were dilatory or rebellious.

If an officer was in the slightest degree disobedient to those above him in rank, he was "broken" as soon as the case was proved to the satisfaction of the principal. If the delinquent was a seaman, under-steward, or fireman, he was relieved from further duty on board, and required to stay on shore under the eye of the instructors, or of Bates, the old salt, who obeyed orders as though they were all written down in the constitution of the State.

As this was the severest punishment that could be inflicted upon any of the students, it soon had its effect. Before the season closed, the ship's company were as obedient to the new officers as they had ever been when Captain Gildrock was in command. More than this, Dory was very popular in the school; he was not unreasonable, snobbish, or tyrannical, and never did violence to the self-respect of any of his shipmates. After they had learned the trick of doing it, it was a pleasure to obey orders.

The students assembled in the boat-house, and all eyes were fixed upon Captain Dornwood, who was to have command of the expedition to Sandy Point, for the operations on shore as well as those on board of the Sylph.

"Perhaps I ought to call for volunteers for the work of to-night, for some of you may not want to sit up so late as the business in hand may require," the captain began, with a cheerful smile on his handsome face, for his good looks had certainly improved in the last two years.

"All night if you like!" shouted Bob Swanton.

"The principal instructed me to say that the early bell will not be rung to-morrow morning," continued Dory. "There will be a good deal of hard work to be done, including some lifting, though there are enough of us to make the task easy. These are the hardships of the trip; and if any student prefers to stay at the school, he will be permitted to do so. If there are any such they will please step forward."

Of course, there was not a single one who wished to be excused from duty. The captain of the Sylph explained that they were to make a trip up the lake by moonlight in the steamer, and do a smart job on the other side. This was all he would tell them at that time, and he directed them to put on their uniform.

CHAPTER VII.

AN EXPEDITION BY MOONLIGHT.

THE Sylph lay at the new wharf, and as soon as the students had put on their uniforms they went on board of her. Chief-engineer Minkfield was directed to get up steam at once. Captain Dornwood ordered one of the quarter boats to be lowered into the water and manned. Taking Thad Glovering, the first officer, with him, he embarked.

At the order of the coxswain the bowman shoved off, and the oars were dropped into the water. The boat was pulled up the little lake to the stone quarries. Mr. Miker, the lessee of the quarries, had made good use of some of the ideas of Bolly Millweed, the architect of the boat-house. The *caisson*, on which the stone posts for the foundations of the structure had been transported, had suggested to him the building of a huge raft, or scow.

He called the craft a "gundalow," which appears to be a corruption of gondola, though the affair bore but little resemblance to the airy boat of the

Venetians. It was fifty feet long and sixteen feet wide. It was decked over and caulked, so that it was as tight as a ship on the ocean. It had a stow-hole at each end; but these compartments were perfectly tight, so that if any water flowed into them it could not get into the large middle chamber upon which the craft depended for its power of flotation.

When heavily loaded with stone, the deck was only a few inches above the level of the water outside. Mr. Miker's principal market for the production of the quarries was at Genverres, though he had sold a large quantity of stone to be delivered in Burlington. In the centre of the deck was a derrick, which was used as a mast when the gundalow went out upon Lake Champlain. She was provided with a large, square sail, but it could be used only when the wind was fair.

On her trips to Genverres she was poled by four or six men, and made very slow progress. But Captain Gildrock had offered Mr. Miker the use of the Sylph to tow her when he wished, for this was nothing but fun to the ship's company, and, as it looked like business to them, they enjoyed it more than mere sailing without a purpose.

The principal made no charge for the use of the

steamer, and Mr. Miker was grateful for the service rendered by the yacht and the students. The gundalow was just the thing Captain Dornwood wanted for the operations of the night. When the boat reached the quarry, the captain went on board and measured it. But the derrick was in the way, and unless it could be removed, the craft would be useless to him.

Returning to the boat, he proceeded farther up the creek, to a point near Mr. Miker's house. Landing again, he found the quarryman in his garden. He stated his business. Of course he could have the use of the gundalow, and the derrick could be taken out of her. The man of stone was enthusiastic to serve the students, and he did not even ask to what use the craft was to be applied, though Dory volunteered the information that the plan he was to carry out was approved by the principal.

Mr. Miker hastened to summon all his men, who lived near the quarries, and by eight o'clock they were on the deck of the gundalow. But it was no small undertaking to remove the derrick, for the mast was a very heavy spar, and was stepped in the bottom of the scow.

The rigging and the long arm were taken from it, and then one of the movable derricks used in the quarries was brought on deck, and guyed up for work. With the aid of this machinery the mast was taken out, and deposited on the shore. The mast-hole was covered with a tight scuttle made for the purpose, and the gundalow was adapted to the business for which she was to be used in the expedition to Sandy Point.

By this time it was nine o'clock, and the moon was just beginning to cast its silvery light upon the still waters of the little lake. Captain Dornwood promised to return the scow to the quarries before morning; but Mr. Miker said he should not use her for a week, and the captain could keep her as long as he wished.

" We shall want a lot of blocks, planks, and timbers, but we have plenty of them on the school grounds, though we shall have to lug them a considerable distance to put them on board of the gundalow," said Captain Dornwood, as he was about to step into the boat.

" Hold on then, Dory! I have everything you can possibly want in that line," interposed Mr. Miker. " The students have saved my men a vast deal of hard work in towing the gundalow, and they will be glad to put all the lumber you need on board of the scow."

" That we will ! " exclaimed several of the men in the same breath.

" I don't want to give you and your men, who have been at work all day, any unnecessary trouble," added Dory.

" No trouble at all ! " protested the men, as they began to put the timbers on board.

Dory was very grateful to them, and pointed out the kind of stuff he wanted, including a large pile of rollers used in moving heavy blocks of stone. In half an hour the gundalow was loaded with the materials Dory had indicated. In the little time at his disposal, the energetic leader of the enterprise had made a list of the material he was likely to require. He had been at work, while the men were loading the blocks and planks, with his pencil and paper, and had thought of several things that were of prime importance.

" I am very much obliged to you, Mr. Miker, and I shall be still more so, if you will lend us eight jack-screws, for we have not enough of them at the shops," continued Dory.

" Are you going to move a meeting-house, Dory ? " asked the quarryman, laughing.

" We are going to do something of that sort," replied the leader of the enterprise. " But I don't let on just yet."

"All right; you know what you are about every time, and it is best to keep your mouth shut, in case you should not succeed as well as you expect. I have a dozen rather small jack-screws, and I will have all of them put on the deck of the gunda-low," added Mr. Miker, as he ordered his men to bring them from a shanty where they were kept under lock and key.

"I will see that everything is brought back again before morning," said Dory, as he stepped into his boat, and gave the order to return to the Sylph.

It was now nearly ten o'clock on as beautiful an evening as ever gladdened the heart of any night wanderers. The full moon gave an abundance of light, and the operations of the students could be as readily conducted as in the day-time. Every-thing that would be needed, with the exception of a few coils of rope, was on board of the gundalow. A party was sent to the shops for them; and when these necessary articles were obtained, the fasts were cast off, and the steamer stood up to the quarries.

The gundalow had been so often towed by the Sylph, that the business was perfectly understood In a few moments more she was made fast to the steamer by the double tow-lines, so that the awkward craft could be steered even around a corner without any difficulty. Will Orwell, the second officer, was detailed to take charge of a party of six on board of the tow. But before the steamer got under way again, Captain Dornwood called all hands together on the forward deck.

"Now we shall know what sort of a racket this is going to be," said Dick Halifax, as they hastened to the place of meeting.

"No, you won't," replied Dick Short, to whom the remark was addressed. "You won't know anything at all about it until we come to the work to be done."

"Why don't he tell us what we are to do?" asked Dick. "I should like to know something about it."

"It was a trick of Captain Gildrock to keep his business to himself, and Dory takes after him. The principal thinks the fellows can obey orders better when they don't know what is coming than they can when they understand all about it. Every fellow thinks he knows best how to do almost anything."

"I don't know but he is right. I never saw a horse tumble down in the street, but every one of the crowd around him wanted to boss the job of getting him on his feet again," added Dick.

" I have called you together, fellows, to say that
it will be necessary to keep as still as possible on
the expedition of to-night;" said Captain Dornwood, when the ship's company had all gathered
on the forward deck. "I don't know that a noise
would defeat our plans, but I am very much afraid
it would cause us some trouble. I don't believe
in any yelling when we are on duty, but I fear it
would make mischief to-night. Please to observe
this request in the strictest possible manner."

" Where are we going, Captain Dornwood?"
asked Bark Duxbury, one of the new students.

" Going to work now," replied the captain with
a smile. "All hands to their stations."

The ship's company separated, and all the officers and seamen went to the places where they
belonged. Though no meals were to be served
during the night, so far as was known, the cooks
went to the galley, and the stewards to the forward
cabin. The second officer, with his gang, went on
board of the gundalow, and at the order from
the captain the pilot on duty rang the bell to back
her. By this movement the scow was hauled out
from the wharf, and the bell to go ahead was given.

Mr. Miker and some of his men stood on the
shore watching the departure of the expedition,

and wondering what sort of a mission the students were going upon at that time in the evening. But the Sylph and her tow soon disappeared beyond the trees at the lower end of Beechwater. Dory was on the hurricane deck, keeping a sharp lookout upon everything that was done.

At the V-point the pilot slowed down without any order from the captain, and the scow was switched around it without touching the mud. There was now nothing to do outside of the engine-room and pilot-house ; and the crew gathered into companies in various parts of the deck to specu-late upon the nature of the expedition in which they were engaged. They guessed a hundred things. The crew of the Goldwing were pretty sure they were going to Sandy Point.

The Sylph was approaching the mouth of the river, and it would soon be necessary for Captain Dornwood to say something. For, if the expedi-tion was bound to the northward, she would take that course as soon as she came up with the point on that side of the river ; if she was going to the southward, she would have to keep her present course half a mile farther out into the lake to avoid the shoals off Field's Bay.

Oscar Chester and Dick Short, the latter of

whom had been promoted from a deck-hand to the position of second pilot, were in the pilot-house. No order came to alter the course at the north point, but a few minutes later the captain entered the pilot-house.

"We are bound to Sandy Point," said he ; and the head of the steamer was turned to the southwest.

In less than half an hour, the Sylph was close in to the end of the point, and Dory discovered Paul on the shore. The steamer was headed into the bay, and the gundalow brought up to a point directly in front of the cottage.

CHAPTER VIII.

A CHANGE OF LOCATION.

BOTH of the quarter boats of the Sylph were lowered into the water, and a shore party landed with Captain Dornwood. The steamer was then left in charge of the first pilot. The hands on board of the gundalow had poled her up to the beach where she had grounded.

"I am glad to see you, Dory," said Paul Bristol, when the captain went on shore. "It was so late that I was afraid you were not coming."

"We have plenty of time to do the job, for I don't think it will take us a great while. Have you seen anything of the Chesterfields this evening?"

"Not one of them has been near the point, so far as I know, and I don't expect to see any of them. I suppose they are dreaming of the fun they will have in pitching the cottage into the lake to-morrow afternoon," added Paul, with a cheerful smile. "But I don't see how you are going to move the building, Dory."

85

"If you keep your eye on us sharp for an hour or so you will see," replied the leader of the enterprise, as he turned his attention to the business before him.

After half an hour's hard work, the lumber, blocks, and rigging on the deck of the scow were landed on the beach. With thirty pairs of hands the work was not very hard, and they tossed the large sticks about as though they had been nothing but chips. By this time they understood what was to be done, and the students were full of enthusiasm. They were required to work in silence; for though the Chesterfield school was all of half a mile from Sandy Point, Dory was very anxious lest their operations should be disturbed by the institute people.

Two heavy timbers were placed under the cottage; the jack-screws were put in position under them, and the building raised from the posts which supported it. A plankway was laid on the smooth sand, the posts were removed, and the cottage set on rollers. The plankway was continued to the water.

There was a considerable descent from the site of the cottage to the water. Two heavy ropes were attached to the building, and passed around a couple of large trees in the rear of it. The plankway was an inclined plane, and it required but little force to start the cottage on its journey. With a couple of turns around the trees, the hands stationed at the check-lines easily controlled its movements, and slacked off only as the captain gave the word.

In a few minutes the building was rolled down almost to the water. The gundalow was aground on the shore end. Two heavy timbers were extended from the deck to the beach and supported by blocks so that they would bear the weight of the structure. These beams lay nearly level when they were in position, and just reached the end of the plankway on shore. The check-lines were eased off again when smooth bearings for the rollers had been prepared.

When the cottage was about half on the timberways the force of gravity was no longer available, and the building refused to budge another inch. While Captain Dornwood was on the front of the structure, some twenty of the students in the rear tried to push it toward the gundalow; but they could not start it.

"Enough of that!" called Dory, as soon as he saw what they were doing. "You are acting without orders, and wasting your strength for nothing."

"But the building sticks fast where it is," said Ben Ludlow.

"If you think you can push it ahead you are mistaken," added the captain. "It has gone as far as I expected it to go of itself."

The two check-lines were then carried on board of the scow, and the Sylph was backed up to her. The lines were made fast at the quarters of the steamer. Dory stood on the after end of the gundalow, and, with a boatswain's whistle, made a signal agreed upon with the pilot to go ahead.

The lines stiffened and strained, and then the cottage began to move again. The timber ways had been continued on the deck of the scow, and the building moved very slowly until the captain gave a second signal with the whistle.

The rollers were instantly blocked by hands under the direction of the first officer. But the rear of the cottage just reached the stern of the gundalow. At least half of the weight of the building rested upon the sand at the bottom. The water deepened very rapidly near the shore on the outside of the point, and it became necessary to handle the heavy burden with the greatest care, for the forward end of the craft would settle down as soon as the structure was moved any further, forming an inclined plane, on which the cottage might roll overboard.

There were four iron rings at the stern of the scow, and check-lines were extended from them to the structure. A double turn was taken in each over a cleat, and hands placed at these ropes. The signal was again given for the steamer to go ahead. The building moved a few feet further, and the rollers were promptly chocked when the captain gave the whistle to " stop her."

The cottage was not yet exactly in the middle of the deck, and another movement was necessary. The bow of the scow settled down, but the check-lines held the house firmly in position. The second move was so well timed that it placed the building in exactly the right place.

The check-lines were belayed under the direction of the first officer, while the second officer proceeded to fasten the cottage to the rings in the bow of the scow. It was to remain on the rollers during the trip to its destination, and Captain Dornwood made sure that it was secured beyond the possibility of any accident.

All the spare hands were then ordered to the

shore, Dory leading the way. The lumber, jack-screws, blocks, and other material were put on the scow, for there was still abundance of space forward and abaft the house. Everything connected with the cottage was put on board.

"By the big wooden spoon!" exclaimed Paul, when the burden of the work was done. "I did n't believe you could do it with a hundred men."

"We have n't finished the job yet," replied Dory, laughing.

"But I believe you can do all the rest of it," added Paul, filled with admiration. "These students are good for something besides keeping bread and meat from spoiling."

"They are good fellows," answered Dory, "but we have not quite finished over here yet."

"You fellows might come over here some night and carry off the building of the Chesterfield Collegiate Institute if you felt like it. I don't see what more there is to do."

A lot of shovels, hoes, and iron rakes had been brought over on the steamer, and these were now carried on shore. The post-holes under the cottage were filled up, every particle of rubbish was removed, and the ground raked over until every thing was as smooth as though no human being had ever resided within a mile of the spot.

"By the big wooden spoon!" shouted Paul. "It looks just as it did when we first came here."

"We will leave everything in good order and condition so that Major Billcord shall have nothing to complain of," replied Dory. "Now make the Dragon fast to the stern of the gundalow, and we will get under way. I think you had better stay in the house to see that everything goes right there."

"All right, Dory, I will do just as you say; but I don't believe you have started a joint in the cottage. I went up to look at the chimney with a lantern while you were shifting it, and there is not a crack in it."

The chimney reached only from a beam to the ridge pole, and a couple of feet above it, so that the brickwork had required no special consideration. But the building had been subjected to no hard usage, and no damage had been done to it. All the furniture remained just as it had been for two years, and Mrs. Bristol might have kept house in it as well as when it was stationary.

As soon as the ship's company were all on board of the steamer, or the scow, the captain gave the word to go ahead. The tow-lines had been adjusted before. The end of the gundalow, which was aground, grated a little on the sand, but it came off without difficulty, and the Sylph with her tow headed down the lake.

The officers of the steamer were so well accustomed to handling the gundalow that no difficulty was experienced in getting the cottage to its destination, which was to be at Hornet Point, near the outlet of the creek into Beechwater. The location had been suggested by Dory, and agreed to by Captain Gildrock. It was quite as pleasant a spot as the former site of the cottage, and was but a short distance from the new boat-house.

The plank and timber ways were laid down as they had been on the other side of the lake, and the building was moved to the shore as readily as it had been put on board of the gundalow. By two o'clock in the morning it was in position on the posts upon which it had rested at Sandy Point. The materials were all conveyed to the quarry, and the gundalow was left at its usual moorings.

By this time most of the students were gaping fearfully, and were very tired. Paul remained at the cottage and went to bed after the departure of the Sylph. The ship's company were dismissed at the wharf, and before half past two they were all asleep in the dormitory. Mrs. Bristol and Lily were up early in the morning, and went out to walk by six o'clock.

After the departure of the students the night before in the steamer, they had not heard a word about the cottage. They walked over to the boat-house, where they found the principal, who was an early riser. The cottage could not be seen from the boat-house, though it could from the wharf. Bates was bringing up a boat in which the captain was going out to inspect the operations of the night.

"Good morning, Mrs. Bristol; good morning, Miss Lily. You are up early," said the principal.

"But I don't see anything of the cottage," replied Mrs. Bristol, after they had returned the pleasant salutations of the captain. "I did not hear a sound in the night, and I suppose Dory was not able to carry out the plans we talked about."

"They certainly did not make any noise about it; but if you and Miss Lily will take a seat in this boat, we shall soon ascertain what has been done," said the captain, as Bates brought one of the four-oar boats to the landing steps.

The ladies seated themselves in the stern-sheets of the boat, and the boatman pulled out into the lake. But he kept near the shore, and the over-

hanging trees obstructed the view of Hornet Point.
In a few minutes, however, the boat was out far
enough to afford its occupants a view of the mouth
of the creek.

"Why, there's the cottage!" exclaimed Lily.
"It looks as though it had stood there since it was
built."

"The boys have done their work very well,"
added Captain Gildrock.

The party landed and walked up to the cottage.
Not a particle of rubbish had been left on the
premises; not a plank or a block. Where the
sand on the beach had been disturbed it had been
raked over, and everything looked as neat as
though the family had lived there for a year.
They went to the front door and the back door,
but both were locked. Paul was still fast asleep
in his chamber, and they did not disturb him.

CHAPTER IX.

THE JANITOR OF THE BOAT-HOUSE.

CAPTAIN GILDROCK was delighted with the skill and the industry which the students had displayed in the removal of the cottage. It was not the difficulty of the feat they had accomplished so much as the neat and orderly as well as quiet manner in which the work had been done. Usually boys cannot do anything without a great noise and not a little bluster. But the Beach Hillers had not disturbed any one on either side of the lake.

With the machinery at their command it was not a great achievement to move a building no larger than the home of the Bristols across the lake. The principal had as yet no report of the work, but, taking the appearance of the cottage at Hornet Point as a specimen of the labor done, nothing could be better.

"Everything seems to be in good order here, Mrs. Bristol," said Captain Gildrock, when he had examined the cottage and its surroundings.

"I can't see for the life of me how the students brought the cottage over here and put it on the posts just as it was before, and in the night, too," added Mrs. Bristol.

"And everything is just as neat as wax-work," said Lily.

"Just beyond the quarries is what we call the lake road, which is the boundary of my land on the east side. There is a driveway from it through the quarries, near the shore of the creek. I shell continue this road to Beechwater, which will carry it by the end of the cottage," continued Captain Gildrock, pointing out the locality. "By this road you can go to the town without passing through the school-grounds, though you are entirely welcome to use the latter."

"You are very kind, sir," replied Mrs. Bristol. "I am sure I have not the slightest claim upon you for anything, and you have done more for me already than all others. We shall be grateful to you as long as we live."

"I think you are a very worthy woman, and I am very glad to be able to serve you," replied the captain. "But I have come to the conclusion that my mission in the world is to help others to help themselves. You have a son and a daughter."

"And they are both able and willing to work," added the woman.

"So I have heard from my nephew; and I expect to put you in the way of earning your living. In the first place what is to be done with your son?"

"He will do any kind of work he can get to do — work in a store or on a farm."

"If he goes into a store, he has about one chance in ten of becoming something more than a counter-jumper on five dollars a week. But he ought to learn a trade."

"I should be very glad to have him do so, but we are dependent upon him just now for the means of living. When Lily had a place in Westport, she received only a dollar a week besides her board; and sometimes Paul could not make any more than that."

"I have a place for Paul. I want a janitor for the boat-house, for Bates is getting rather too old to do such work. I will give your son a salary of twenty dollars a month for the service."

"You are very kind, sir; that is more than we ever had to live on," replied Mrs. Bristol.

"But I think he had better join the school at the same time. We can make a carpenter or a machinist of him; and if he prefers some other trade, what he learns here will not come amiss. He can do his work in the boat-house and be a member of the school at the same time, though he will have to work some part of the day while the students are at play."

"Paul will be very glad to work and never play, for he has always been a very good boy," added the devoted mother.

"Your daughter, you said, had worked at the millinery business, and perhaps a place can be found for her in Genverres," continued the captain, as he led the way back to the boat. "We will go to breakfast now."

The family took their morning meal at the usual hour; but not a single student had yet appeared on the grounds. The principal would not allow them to be disturbed until nine o'clock, when the bell was rung in the dormitory, though a few of the boys had turned out at this hour. At half-past nine breakfast was served to them; and they all appeared to be in as good condition as usual.

Paul was invited to join them, though he was to board at home as soon as his mother was established in the cottage. By this time he was pretty well acquainted with the students, and was very

popular among them.   The story of his fight with
Walk Billcord and his father had been told on
board of the Sylph on the passage to Sandy Point,
and his prowess made him a hero among the boys.

"Paul, did you take the tin box from the hollow
of the tree?" asked Mrs. Bristol, as her son was
leaving the mansion with the rest of the students,
for Fatima Millweed had already entered his name
on the roll.

"I never thought a word about it, mother,"
replied Paul, not a little mortified at the neglect.
"I was so busy and so anxious that it never came
into my head.   But I will go over in the Dragon
and get it right off."

"But Captain Gildrock has a place for you as
janitor of the boat-house, and perhaps he cannot
spare you," added Mrs. Bristol.

"Janitor of the boat-house!" exclaimed Paul,
opening his eyes very wide.

"And your salary is to be twenty dollars a
month," continued Mrs. Bristol.   "The captain
says his mission is to help those who are willing
to help themselves.   Besides this, you are to be
one of the students, and learn to be a carpenter
or a machinist."

"One of the students!" almost screamed Paul.

"But you will have to work while the other students play, my son."

" By the big wooden spoon ! I shall be willing to work all night if I can learn what the other fellows learn," replied Paul.

His mother explained to him more fully the intentions of the principal, and the son of toil was more delighted than if a fortune had suddenly dropped into his lap. He knew all about the course of study at Beech Hill, and thought it was the finest school in the world. He had long wished that he might learn a trade, and he would have sought a place with a carpenter before, but he would have to work for nothing at first, and his mother needed the dollar or two a week he could earn.

" When will Paul begin his work as janitor? " asked Mrs. Bristol, as the principal was passing them in the hall.

" At once, Mrs. Bristol. His wages shall begin to-day," replied Captain Gildrock. " But if you want him at the cottage till you get settled we can spare him, though he had better join his class to-day."

" I wanted him to go over to Sandy Point," continued Mrs. Bristol, who then explained the

errand upon which she proposed to send Paul.
"The tin box contains my wedding ring, my
mother's gold ring, and a two-dollar bill. I was
afraid to keep them in the house, for rough characters sometimes land at the point. I did n't think
of the box till I wanted the money to buy some
provisions."

"But Paul would have to row ten miles to get
the box," added the principal. "This is a broken
day, and we shall not do much in the school or the
shops, and he can go over in the Goldwing after
the students are dismissed. I will pay Paul's
first month's wages in advance, for I am sure you
will want some money."

The good woman took the money under protest,
though it was true that she needed it. The gold
rings were of more value to her than any sum of
money, and she hoped they would not be lost.

At ten o'clock the bell rung for the forenoon
studies. Paul took a desk assigned to him, and no
other boy was ever more interested in a circus
than he was in the exercises of the school-room.
As soon as the school was assembled, Captain
Gildrock took the platform and called upon the
leader of the moonlight expedition to report upon
the action he had taken.

Dory modestly related all the particulars of the
trip to Sandy Point, and the removal of the cottage, and warmly commended the ship's company
for the good order they had maintained, the
promptness with which his orders had been
obeyed, and the quietness with which all had done
their duty.

The principal believed in giving reasonable com
mendation when it was deserved, and he bestowed
handsome praise upon them on this occasion.

When the boys came into the school-room, they
noticed upon the wall in the rear of the platform
a large drawing which they had never seen before.
It consisted of three plans of a vessel. On a table
was a model of the hull of a craft of some sort,
resting in a cradle.   The students had kept their
eyes fixed on the drawings and the model most of
the time while they listened to the commendation
of the principal and the report of their leader.

They manifested a very strong interest in these
things, and they were likely soon to forget the
operations of the night before. For six months
there had been a great deal of talk among them
about building a boat, and the project was a very
attractive one to them. But up to the present
time nothing had officially been said or done about

it. As soon as the spring opened, they had been required to erect a sort of shop on the very bank of the little lake, near the old wharf of the steamer.

This structure was seventy-five feet long, with plenty of windows, and was entirely open on the water side. In accordance with the general policy of the principal, its use had not been explained; but all the students believed it was to contain the ways on which the boat was to be built. It looked now as though the desired information in regard to the building of the boat was to be communicated to them.

"I need not ask you if you have noticed these drawings, and this model of a vessel," said Captain Gildrock, after he had finished what he had to say about the moving of the cottage; "for you have been looking at these things most of the time since you came into the school-room."

"Are we to begin on the boat to-day?" Lon Dorset asked; and he was one of the new students, not yet thoroughly broken in with the customs of the school.

"When I set you at work you will begin; not before. It always affords me very great pleasure to answer sensible questions, boys, and I shall do everything I can to encourage you to ask them; but I don't believe in foolish questions. Such is the character of all questions relating to what we are going to do. You are never required to do anything until an order is given. Foolish questions take up as much time as sensible ones."

Lon Dorset was somewhat abashed at the manner in which his inquiry had been treated; but the principal knew that some of the boys would talk all day about nothing, if permitted to do so; and the questions he tolerated and encouraged were those which brought out real information, and revealed the condition of the inquirer's mind.

" The building of the boat has been somewhat delayed on account of the difficulty of obtaining suitable lumber," continued the principal. " A load which came from Boston yesterday will enable us to make a beginning."

Some of the new pupils were disposed to give three cheers.

CHAPTER X.

W E are not ship-builders, boys ; in fact, there is not a ship-builder connected with the school, and I do not intend to engage one even as an instructor," said the principal, continuing his remarks on the platform. "In the present depressed state of this important industry, perhaps it is not advisable to devote much time to the study of scientific construction in ship-building. It looks now as though the ships of the future were to be of iron ; and many vessels of this material are built in this country at the present time.

"But perhaps ship-building is rather too ambitious a term to apply to our intended operations. We shall build a boat of considerable size, and while we are doing the work we shall learn what we can about ship-building. Many years ago I built a ship for myself, and superintended its construction from the keel to the trucks. In building our boat we shall not put in every stick used in a ship.

"Did any of you ever tow a log in the water?' asked the principal, pausing for a reply.

"I have," answered Leo Pownall, whose father owned a saw-mill. "I have towed lots of them on the mill-pond."

"To which end of the log did you make fast?" inquired the captain.

"To either end; just as it came handy," replied the student.

"Then you sometimes did more work than was necessary with your oars. A log tows easier when you make fast to the big end," continued the principal, waiting for the pupils to digest the idea.

"I don't see what difference it can make," added Leo. "If anything, I should say that the small end would open a passage through the water more readily than the big end."

"I suppose none of you ever saw a whale, but most of you have caught horn-pouts, or bull-heads."

"I have seen a whale on exhibition in New York," interposed Luke Bennington.

"What was the shape of his head?"

"The one I saw was round; but I have seen pictures of whales in which the head was nearly square."

"How is it with the pout!"

"His mouth is about the widest part of him," laughed Alick Hartford.

"Take fishes in general, in what part of the body do you find the greatest girt?" asked the principal.

"Just astern of the head," replied Kit Burlington.

"In some fishes, about one third of the length from the mouth," added Bark Duxbury.

"Very good; you are about right, though some fishes vary from the general rule. Now don't you think Nature made a mistake, Leo Pownall, and that fishes ought to swim tail first instead of head first, as you would tow a log?"

"I suppose God made the fishes all right; but He gave some of them very sharp noses," returned the saw-mill owner's son.

"Corresponding to the shape of the butt of a log after it has been felled; but the greatest girt is still near the head. This is the general shape of the hull of a vessel."

"But the head of a sperm whale is almost square; and no other fish is like him," added Phil Gawner.

"The whale is not a fish, Gawner. I have seen

a school of porpoises alongside an ocean steamer. Their greatest girt is one third of the length from the head end; but they will swim past a fast steamer, and make something like twenty knots an hour," said Captain Gildrock.

"I was trying to find the porpoise in Wood's Natural History the other day; but there is no such fish in the book," added Sol Guilford.

"Where did you look?"

"In the volume about fishes."

"The porpoise is not a fish, and you would have found it in the volume marked 'Mammalia,'" replied the principal with a smile.

"But is n't the porpoise a fish? He lives in the water."

"So do hippopotami; but they are not fishes. Whales, porpoises, dolphins, seals, and some others, are mammals; that is, they suckle their young as a cow does a calf. Properly they are not fishes, though they are very often called so."

These were the kind of questions the captain believed in encouraging, though they sometimes led the conversation out of the legitimate channel. They elicited useful information; and he was careful not to let the students wander too wide of the subject under discussion.

"I don't know now why a log or a fish goes best with the big end ahead," said Leo Pownall.

"After the passage for a moving body in the water is opened, this fluid follows its own laws, and seeks an equilibrium. As it moves back to its natural level, it crowds in upon the after part of the body, whether it be a log, a fish, or a ship, and thus pushes it ahead. Under the stern of a vessel, the hull is curved, or hollowed out, just as the size of a fish diminishes at the tail, which is the fish's rudder.

"But the shape of the hull is varied according to the use to which the vessel is to be put; but the rule will hold good in the main. In building a ship the beginning of the work is done on paper. As in erecting a house, the first thing is to obtain the plans, which are made by the naval architect. In fact, the entire shape of the vessel is laid down on the drawing-board. From these the builder gets his dimensions, all the curves, and the form of every timber and piece of wood used.

"On the drawing on the wall," continued the principal, taking the pointer and indicating the plans, "everything is put down that can be needed in the construction of the boat we intend to build. There are three plans, you will observe. I had

them drawn by a naval architect in New York.
This," and the principal pointed to the highest one
on the paper, " is the sheer plan.   It shows the
side or profile of the hull on a flat surface.   It
looks just as the broadside of the Sylph would, if
she were too far off for you to get any idea of the
curves in her sides.

    " This plan gives the exact curve of the bow,
and the exact slant of the stern-post.   The three
straight lines extending the whole length of the
hull are the levels to which the water would stand
if the vessel were submerged to three different
depths.   This drawing is made on a scale of one
inch to a foot.   The sheer plan is a vertical plane
through the keel.   From it we get the length and
height.   The red lines which extend from certain
points at the bow to the lower part of the stern
post indicate the various curves of the hull at
different distances from the vertical plane of the
keel.   In other words, they are three vertical
planes, parallel with the central plane.

    " The next plan, of the same length as the first,
shows you one half of the deck of the boat, and
is called the half-breadth plan.   All the plans are
on the same scale.   The straight lines on the deck
are the curved lines of the sheer plan, or the

tops of the several vertical planes. This plan re-
versed would show the other side of the vessel.

"The third is the body plan, and exhibits a vertical section of the hull, looking at it end-on, at
the point where it has the greatest breadth. The
right-hand half of it shows the bows, and the left
the stern. The curved lines are the same as those
on the sheer plan, though, of course, they are shown
only at the bow and stern, for you cannot see a
line when you look end-on.

"With the making of these plans the task of the
naval architect comes to an end, unless he is employed to superintend the construction of the
vessel. From the plans the builder gets the exact
size and shape of the craft he is to build. From
it the moulds, or patterns, of all parts of the hull
are made. In an apartment called the moulding-
room, with which every ship-yard is provided,
full-sized plans of the vessel are drawn on the
floor. I do not mean that the entire ship is drawn
at the same time.

"If the bow or stern was accurately transferred to the floor, enlarged to the actual size of
the hull, the exact form of the stem or stern post
could be marked off. From this, a mould or pattern could be made of board or plank. As a

matter of fact, a mould is made for every part used in the construction of the ship, not every piece of wood, for what is used for one side may do just as well for the other side. For example, a timber on one side is exactly like the one on the opposite side.

"In ship-building, the word timber has two meanings. As in general use, it may be any large stick of wood. In the technical sense, it is one of the ribs of the vessel. The means of understanding which is meant will be given you as you proceed with the work. The keel is the backbone of the vessel, and the strength of the hull depends largely upon it.

"The keel corresponds with the backbone of your bodies. At the forward end of it is the stem, shaped as you see in the sheer plan. At the after end is the stern-post: and these three parts form the profile of the vessel. Between them are the timbers, or ribs, curved as required by the shape of the hull. To the timbers, the stem and stern-post, the planking, or outer skin of the ship, is fastened. If you were to build a canvas canoe, you would make a frame such as I have described. The cloth part would correspond to the planking.

"But, besides the principal parts I have mentioned, of course there is a vast number of other parts, whose names you could not remember if I gave them to you. I shall add only a few of the principal ones. The timbers — I use the word in its technical sense — are set up about three feet apart, sometimes a little more, and sometimes a little less. The lower end of each is fastened to the keel, and of course each timber has to be shored up, and carefully secured in its proper position.

"The timbers are at right angles with the keel, and in large vessels are formed of several pieces. Across the keel is laid the floor timber, which is the connecting link between the pair of ribs. In the middle of the ship, the floor is nearly flat; but near the stern-post the timbers strike the keel at an acute angle, and in the same manner at a less angle at the bows.

"Above the floor timbers is laid the keelson, which is a large and strong timber a foot square or more in large vessels. These pieces are fitted together, and bolted to the keel through the timbers. The sharp angle at the stern is filled with solid wood. As you have seen, the floor timbers are parts of the ribs," continued the principal, pointing to a diagram of a full rib. " The next two sticks, forming the sharpest bend in the timber, are the futtocks, above which are the top timbers and the lengthening pieces. The plankshear is placed on the top of the timbers, extending from bow to stern, even with the upper deck, if there is more than one.

" Large ships have two, three, and even four decks. Fastened to the timbers are pieces called shelves, upon which rest the beams or timbers extending across the ship, upon which the decks are laid. In the corners, where the beams join the ribs, are placed the knees (timbers like brackets) in which the angle is formed by the natural growth of the wood. Of course all these pieces are bolted together in the strongest manner.

" The timbers next to the stem are the knight-heads. They extend upwards higher than the tops of the other ribs, and assist in the support of the bowsprit. In the keel, stem, and stern-post, a rabbet, or triangular groove, is cut out, into which the planks are extended. The first course of the planking, next to the keel, whether composed of one or many pieces, is called the garboard streak. This word is often written *strake*. The other word is more commonly used in this country.

"The next course above the garboard streak is the bilge streak, which may consist of several widths of plank. Above there are the wales, and still higher the shear streaks. Some of these terms are applied to the parts of the ship as localities. The bilge is where the sharpest bend comes in the hull; the wales are the sides near the load line.

"I have tried to give you a general view of ship-building, with a few of the more important technical terms, some of which most of you have learned before. As I have said, we shall not use all these pieces in building the boat. For example, a false keel is put under the true keel of a ship. It is a timber of the toughest wood, from four to six inches thick, which is bolted to the keel after the keelson is secured. It is but lightly fastened, for it is intended to come off, if the vessel strikes a shoal, and thus allow the true keel to slide off. We shall not need this addition, unless we require it to increase the depth of the keel. In that case, it would be better to have the part corresponding to the false keel made of lead or iron, and then it will serve as so much ballast.

"On the table you see a wooden model of the boat we are to build. Its form and size are exactly indicated by the three plans I have explained. We are not ship-builders, only amateurs; and, while I shall take pains to have you understand the theory and practice of the art, I do not feel obliged to follow all the methods in use. So far as I know, no such model as the one on the table was ever made before. As I shall direct the construction of the boat, I shall do it in my own way, though it may not be according to the accepted rules.

"I have kept you now longer than I intended, for, after the hard work you did last night, and the very quiet and business-like way in which you did it, I shall make the rest of the day a holiday. The Beech Hill fleet is at your service, and you may spend the day in any proper manner that you please. To-morrow afternoon we will dissect this model, and give out the work of building the boat. In the meantime I shall be glad to receive suggestions as to her name; but no student must send in more than one name, for I wish you to have decided opinions."

CHAPTER XI.

ROUGH WATER ON LAKE CHAMPLAIN.

THE wind was very fresh from the northwest on the lake, and its surface was covered with white-caps. Above Split Rock Point the lake looked like a sea of foam, and heavy waves rolled in upon the beach in Porter's Bay. Even Beechwater was considerably agitated. Prudent mothers would have thought it dangerous to go out upon the lake at such a time.

Apparently for the reason that the water was rough on Lake Champlain, the students elected to have an excursion in the barges. The principal did not object, for the boys had been trained to rough weather, and they knew how to handle the boats in any sea that ever was stirred up on fresh water, albeit the waves are often more perilous on large lakes than on the ocean.

Oscar Chester was the coxswain of the Gildrock, and Thad Glovering of the Winooski. The former was still used by the first class, and the latter by

the second.   But the classes had been considerably
changed, and vacancies in the first had been filled
from the second and from the most advanced of
the new pupils.   The third class consisted mainly
of new scholars.

The twelve-oar barges each had a crew of thirteen, including the coxswains.   Nine of the third
class were detailed for service in the new eight-oar
barge, and there was one who had no station in
any boat.   One of the thirty-six students to which
the school was now limited had been taken sick,
and returned to his home in the winter.   He was
from the far South, and the climate was too severe
for him.   His place had not been filled before the
coming of Paul Bristol.   He was to be a spare
hand for the present, and was to take the place of
any one who was absent.

The eight-oar barge was the Marian, in honor of
Dory's sister, and the name had been given by the
students.   Paul had spoken to Dory about the tin
box in the hollow of the tree, and it had been
arranged to visit Sandy Point in the Goldwing;
but when the students decided to go out in the
barges, the plan had been changed.   Dick Short,
though a first-rate mechanic, and one of the best
boatmen in the fleet, had been transferred to the

third class because he was deficient in some of his book studies, and could not keep up with his class-mates.

Dick had been elected coxswain in the Marian. Dory had trained the new crew, but he declined to be the chief in the boat. He pulled the stroke oar, though he exchanged places with the coxswain when the boys were in training. The crew of the Marian generally manned the Goldwing, though the schooner was often used by other parties.

A lunch had been put up for each of the crews of the barges, and they were expected to be absent all the rest of the day. Sometimes Captain Gildrock was called by business or pleasure to visit Burlington, Plattsburgh, or other places on the lake, when the students were at their studies, or off in the barges. At such times he was his own pilot, Mr. Jepson was the engineer when not instructing in the shop or drawing-room, Bates was deck-hand, and Collins, the gardener, was the fireman.

Before the students were dismissed from the school-room, steam was up on the Sylph, and the barges had hardly departed before she left the wharf. If she had any particular destination, it was unknown to most of the students; and possibly the principal desired to take a view of Sandy Point after the event of the preceding night.

It was half-past eleven when the barges backed out of the boat-house. Paul was seated in the stern-sheets of the Marian. He had not yet been trained to pull with the crew, though he would have gladly taken an oar. On the present occasion his mission related to business. While they were at breakfast, Lily had spoken to him about a valise she had been obliged to leave at the house of her former employer in Westport. Paul had promised to get it when he could.

The Marian led the way down the creek into the river, and then out into the lake. The other barges followed at a respectful distance, and their crews did not seem to be inclined to engage in any scrub races. The speed of the eight-oar boat had not yet been tested, and it had always been taken for granted that either of the other barges could beat her without half trying. But it was no time to indulge in a race when the water was so rough.

The waves were dashing smartly over the point at the mouth of Beaver River, and the bow of the Marian was lifted up in the air as she plunged in among the white-caps. Dick Short got the hang of the waves as soon as they struck the boat. Paul

thought it was about as rough a time as he had
ever seen on the lake during the season of navigation; but he had never been in the barge before,
and everything was new to him.

"By the big wooden spoon!" exclaimed the
passenger, when the Marian was in the thickest of
the miniature billows, and the water was occasionally slopping in over the bow. "Don't you expect
you will all get drowned?"

"You can't tell about that," replied the coxswain, who felt as much at home in the boat as he
would in the school-room. "We are not prophets,
and we can't tell what is going to happen."

"Don't you think it is dangerous to come out
here when the lake is boiling after this sort?"
asked Paul, as he looked at the angry waves
around him.

"I suppose it is. There is always water enough
in the lake to drown the whole of us," answered
Dick Short, who was rather inclined to work upon
the fears of a timid voyager.

"Then what do you come out here for?"

"For fun."

"Is there any fun in being drowned, Dick
Short?" asked Paul seriously, as he glanced at
Dory, whose face was as calm as the minister's on
Sunday.

"Any fun in being drowned? How should I know? I never tried it," returned the coxswain.

"But don't you think it is dangerous to be out here in such a blow?" Paul insisted; and he really believed he was in peril.

"Of course it is."

"Then don't you think you had better put back into the river?"

"But it is dangerous in there," added Dick. "Suppose a tree should blow down and kill every fellow in the boat? Suppose the sea-serpent should be having a vacation up Beaver River, and take it into his head to swallow us all, one at a time? Suppose the river should catch fire and burn us all up? Suppose the sky should fall, as Chicken Little said it would, and smash us all to jelly?"

"You are making fun of me, Dick," said Paul, laughing.

"The principal says it is useless to worry about anything. We do the best we can with the boat; and if she spills us into the fluid, all we have to do is to get out if we can."

"I think I can stand it as well as any of the rest of the fellows can, and I don't mean to worry," returned Paul. "I never was out on the lake when it was anything near as rough as it is to-day, and it looks dangerous to me."

"If you don't feel right about it, we will put you ashore," added the coxswain.

"I am not scared; I can stand it as well as the rest of you. I was only asking about it for information," continued Paul.

"I don't believe you are scared; if you had pluck enough to stand up against Walk Billcord and his father, I don't believe you will mind a little ripple on the lake like this," replied Dick, laughing.

"You said it was dangerous."

"Well, an ox-team is dangerous if you let the beasts run off a precipice. It is dangerous to go to bed, for the house may burn up before morning."

"We don't think we are in any more danger here, Paul, than we are every hour of the day on shore," added Dory. "Dick could upset the boat, and spill us all into the drink, if he did not understand his business and attend to it."

"All right; I am satisfied," replied the son of toil. "But I thought you were going up to Sandy Point."

"So we are; but the direct course would be

about southwest, and that would put us into the
trough of the sea and keep us rolling the gunwale
under all the time," replied Dick. "The principal don't allow the fellows to be reckless. There
comes the Sylph out of the river, and Captain
Gildrock is on board of her. If he should see
me letting the barge wallow about in the trough
of the sea, when there is no need of it, he would
give me fits."

"We could go direct to Sandy Point, though
at a little risk. We should take in a good deal
of water, and it would be uncomfortable," said
Dory. "In a small boat in a blow, or in a squall,
the safe way is to keep her head up to the sea."

"The other boats are following us."

"The coxswain of each can do as he pleases;
but the fellows all know there is no fun in being
knocked about in the trough of a smart sea,"
added the coxswain. "Do you expect to find any
one at the point when we get there, Paul?"

"No; it is n't twelve o'clock yet. The Chesterfields are in school from eight to one, and then go
to dinner. They won't get away from the house
before two," replied Paul. "I don't believe any
one at the school has any idea of what was done at
the point last night."

"I should like to be where I could see them when they get to the point, and find that the cottage is missing," added Dick, chuckling. "Major Billcord will be the maddest man in the State of New York when he finds it is gone."

"Of course he will be. He don't care anything about the land over there, and all he wanted was to punish us for resisting his saintly son."

"I don't believe it will be safe for you to show your head in Westport again, Paul, or let any of the Chesterfields see you."

"Land me in Westport and see," laughed Paul. Dick agreed to do so.

CHAPTER XII.

A SAILBOAT IN THE TROUGH OF THE SEA.

WHEN the Marian was half-way across the lake, the waves began to diminish in force; and within an eighth of a mile of the high shore the water was comparatively smooth. The barge was then headed to the southwest, and had a quiet time of it till she reached Sandy Point. The Gildrock and the Winooski had followed her, and were now about an eighth of a mile astern of her.

Dory Dornwood was very popular with all the students, not because he was the nephew of the principal, but on account of his fairness, his pluck, and his good judgment. Though Captain Gildrock believed and trusted in him, no one could accuse him of partiality. Perhaps the coxswains of the two twelve-oar barges, who knew that Dory was on board of the Marian, considered it wise and prudent to follow the lead of the eight-oar barge for this reason.

On shore everything was as silent as the tomb.

At Sandy Point, Paul looked with deep interest for the appearance of any person in the vicinity of the site where the cottage had stood. It was possible that Major Billcord had sent one of his men from Westport to ascertain what the Bristol family intended to do about the removal of the cottage or the furniture which it contained; but Paul could see no one.

"It looks as though the coast was clear," said he, when he had completed his survey of the point and the woods in the rear. "I don't believe any one has been here since we left last night."

"Major Billcord must have regarded it as utterly impossible for your mother or you to do anything more than remove some of your furniture," added Dory. "I am sure he did not think of such a thing as your taking the cottage away; and I don't believe he would have considered it possible for the Beech Hillers to do such a job. Probably he did not count us in, or think of us at all."

"It was lucky for my mother that you came along in the Goldwing as you did, for you have saved her all she had in the world," said Paul, with enthusiasm.

"Now, where is the tin box in the hollow of a

tree?" asked Dick Short, as the barge approached the entrance to Sandy Bay.

"The tree is near the neck, and I had to climb up about ten feet to reach the hollow in which the tin box was put," replied Paul. "As the Chesterfields are expecting to have a big time in dumping the cottage into the lake, this afternoon, they may come up early. I have no doubt they will pull around here in their boats."

"Then I think we had better get away from the point as soon as possible," replied the coxswain. "We don't want to get into any row with them."

"I suppose you are not afraid of them," added Paul, laughing.

"I don't think we are, and most of the fellows wouldn't enjoy anything better than a skirmish with them," replied Dick Short. "But the student that does anything to bring on a row with them would be out of favor with the principal, and might have to spend a few days in the brig for it."

Paul had never heard of the brig, and Dick described the strong-room, or black hole, to him. The brig is the place of confinement, or prison, on board ships of war, and the principal had such an apartment in the dormitory. But there had been very little use for it since the earlier days of the school, and not half a dozen of the students had ever seen the inside of it.

"I don't see any of the Chesterfield boats," added Paul, as he looked along the shore. "By the big wooden spoon! Isn't the lake stirred up ahead of us!"

"The wind has full sweep across North West Bay, where the lake is four miles wide. It looks decidedly foamy over in Button Bay," replied Dick Short.

"By the big wooden spoon!" repeated Paul, as he rose in his seat in the stern-sheets.

"Sit down, Paul," said the coxswain, rather sharply. "We don't allow any fellow to stand up in this boat when he gets excited. What is the matter now?"

"There is a sailboat over there, and she looks as though she was tipping over!" exclaimed Paul, dropping into his seat.

"She is over, as true as you live," added Dick, rather louder than he usually spoke, but with hardly more excitement, so thoroughly had the students been trained to keep cool in emergencies.

At the same time he glanced at his crew; but not one of them had turned around to obtain a view of the event described by Paul and the coxswain, for they had been schooled to keep their eyes on the officer of the boat. The crew took more pride in observing this general order than almost any other.

Dick Short gazed with all his might at the struggling sailboat, for a moment, but he seemed to be in doubt, for the craft was at least a mile distant. Besides himself, no one but Paul, whose judgment in regard to the management of a sailboat was not to be relied upon, had even glanced in the direction indicated.

"Stand by to toss!" called Dick. "Toss!"

At the last word the crew brought their oars to a perpendicular.

"Now you can look, and I wish you would do so," continued the coxswain, as he fixed his own gaze upon the sail, which was dead to leeward, and some distance south of Button Island.

The students were glad enough of the permission, for they had as much curiosity, and were as much disposed to get excited, as the average of boys. They gazed with all their eyes at the sail in the distance.

"What do you think of it, Dory?" asked Dick Short.

"I should say that sailboat is half full of water, and that the skipper has lost his head," replied Dory, after he had taken in the situation. "She is rolling in the trough of the sea, and they seem to be trying to take in sail."

All the crew gazed in silence at the sailboat; but no one of them ventured to give an opinion, if he had any, in relation to the disaster. Dory had more experience in sailing a boat than any other student, and perhaps they were not inclined to speak in the presence of an expert. But Dick Short was an excellent boatman, and he deferred only to the skipper of the Goldwing.

"She must be rolling the water into her all the time, and she may go to the bottom at any moment," added the coxswain, whose opinion coincided with that of Dory. "We must go to their assistance at once."

Dory indicated his assent to this proposition only by a nod of his head, for he did not like to appear before the crew to be even an adviser of the coxswain.

"Ready!" called Dick; at which every member of the crew at the oars fixed his eyes upon the officer.

"Let fall!" and all the blades dropped into the

water. "Give way!" and the rowers bent to their oars.

The Marian was headed towards the disabled sailboat, and in a few moments she was going at full speed. The coxswain did not hurry the oarsmen, for he knew better than to exhaust them before the hard work came on. The lake was comparatively smooth under the lee of the land, but in a few minutes they would be in the boiling waves of the broad bay.

"Have you seen anything of the Sylph?" asked Dory of the coxswain.

"She went up the lake when we crossed to the west shore," replied Dick. "The last I saw of her she was off Scotch Bonnet. I think the principal has gone up to Port Henry to order a barge-load of coal, for I heard him tell Mr. Jepson he should do so soon."

"Then by this time he is too far off to see that sailboat," added Dory.

"He couldn't do much if he did see it, for he has not hands enough to handle the steamer and man a boat," said Dick.

"He would manage to render all the assistance needed if he saw the boat," replied Dory, with a smile; for he could not conceive of such a thing as his uncle failing in any duty in an emergency. "He could put the sailboat under the lee of the Sylph, and take every person out of her."

"Of course he would do all he could, and he would save the people at all hazards," continued Dick, still straining his vision to get a better idea of the situation of the sailboat. "But how about the tin box in the hollow of the tree, Paul?"

"I shall have time enough to get that before the fellows go to the point to tip the cottage over into the lake," replied Paul. "This boat begins to leap like a greyhound chasing a rabbit."

"The boat will do very well as long as we can keep her end-on to the sea," added Dory, who thought the new pupil might be alarmed when the barge got into the worst of it. "But remember that you are to do nothing without orders from the coxswain. Simply keep your seat and look out for yourself."

"I think I can stand it as long as the rest of you," replied Paul, with a cheerful smile. "I won't meddle with anything till I am told to do so."

"The Gildrock and the Winooski are following us, and the fellows are putting in the heavy strokes," said Dory.

"Are they gaining on us?" asked Dick.

"I think not."

The sea was very heavy ahead of the Marian, but the waves were not like those of the ocean. They were shorter and more "choppy." But the boats made tolerably good weather among them. In a smart sea, speed is desirable; and it is the element in the progress of the boat which insures safety. At such a time there are two forces acting, the propelling power of the boat and the action of the waves. In heavy weather there is a struggle between the two forces. In the case of the sailing craft, the waves had got the better of the boat.

With the three barges, the advantage was on the side of the boats. They went ahead fast enough to keep the upper hand of the waves.

CHAPTER XIII.

THE stout students at the oars of the Marian drove the barge ahead, helped somewhat by the wind, so that the great billows seemed to have no effect upon her. In a few minutes she was in the midst of the heaviest of the waves. Sometimes she trembled and shook, but she did not yield sensibly to the power which was opposed to her

"I think that is Tom Bissell's boat," said Paul Bristol, who was watching the craft in trouble very attentively. "When I went to see my sister in Westport, about a month ago, she was sewing a full moon into a blue flag."

"A full moon?" queried Dick.

"It was a round piece of white stuff, and it looked like a full moon."

"She has a burgee with a white circle on a blue ground," added Dick. "Then that must be Tom Bissell's boat?"

"She is a sloop as big as the Goldwing," continued Paul.

" Who is Tom Bissell? " asked the coxswain.

" He runs a store in Westport, and his wife keeps a millinery shop in the same building. My sister worked for them," replied Paul.

" Does he know how to handle a sailboat? "

" He thinks he does, and most people believe he does."

" Perhaps he does, and has met with some accident to his sail or rigging," added Dick. " I believe there are some ladies in the boat."

" I shouldn't wonder, for his wife is as fond of sailing as he is ; and sometimes he takes out the girls that work for them," said Paul.

" The sloop is in the trough of the sea, rolling very badly. She is having a rough time of it," continued Dick, as the Marian came near enough for him to see the position of the craft.

" I am very sure that is the Silver Moon," added Paul.

" If it is, it looks like a wet moon, as Bates calls it, when he looks to the silvery orb for the state of the weather," said Dick.

The waves were certainly having it all their own way so far as the Silver Moon was concerned. The peak of the mainsail had been dropped, and the main sheet had run out so that the boom stood at

"The craft contained six ladies and one man." — Page 137.

right angles with the keel. Either the halyards were foul, or the rudder had given out, or she was suffering from both of these mishaps. As the sloop rolled with a heavy jerk in the violent seas, the ladies screamed as though they expected each movement would send them to the bottom.

The craft contained six ladies and one man. The latter was baling out the boat with a bucket, and was working with all his might. He had pluck enough; but the sloop seemed to be dipping up more water than he could possibly throw out, though he had doubtless delayed a little the catastrophe which awaited him.

Dick looked back at the other barges, and he was confident that they had not gained even a length upon the Marian since they all started for the craft in peril. He had not driven his crew, though he had kept them at work briskly. In the barge fleet, Oscar Chester, as coxswain of the senior boat, as the Gildrock was ranked, was the acting commodore. When the barges were within hail of each other, he was in command of the fleet. If the Winooski and Marian were within hail, the command devolved upon Thad Glovering.

As long as the other barges were as far off as at present, Dick Short was in full command. He could

use such measures as he thought best, but the coxswain of either of the other boats could take the
management of the affair into his own hands, if he
chose to do so. As the Marian approached the
sloop, Dick quickened the movements of his crew,
for he desired to take some action before he was
superseded in the command.

The Silver Moon was headed to the northeast,
and lay in the trough of the sea. She was rolling
like a round log in the heavy waves. She had
settled down deep in the water, and behaved like
a stick of wood. The skipper was doing nothing
at all to combat with the waves. As there was no
power exerted to force the boat ahead, she had no
steerage way, and the rudder was as useless as the
spare tiller.

The Marian went as closely astern of the Silver
Moon as she could without fouling the port oars.
The moment they were clear of the hull of the
sloop it was time to execute the difficult manœuvre
of the occasion. In coming entirely about it was
necessary to put the barge in the trough of the sea
for an instant, and this was the dangerous point.

But Dick Short had decided to pass this point of
danger as nearly under the lee of the Silver Moon
as he could. The water on the starboard of the

sloop was a trifle smoother for a couple of fathoms. It required a nice measurement of distances with the eye to handle the boat, and a prompt obedience of orders on the part of the crew.

"Port side! Stand by to lay on your oars!" called the coxswain, when the Marian was astern of the sloop. "Oars!" he added, as soon as the last blade on the port was clear of the hull. "On the starboard, give way lively!"

The effect of the first stroke of the starboard oars, after the order was given, was to throw the head of the barge to port. A few more pulls brought the boat into the trough of the sea; but it remained in that position only an instant.

"Port oars!" continued Dick. "Hold water! Stern all!"

The port oarsmen backed water as the starboard rowers gathered up their blades, so that no confusion occurred, and in less than half a minute the Marian was headed up to the sea, with her stem within a few feet of the sloop.

"On the port, oars!" At this command, the oarsmen indicated lay upon their oars again, and seemed as unmoved as though they had been in the school-room, and not one of them looked behind him.

All the crew had obtained a single glance at the
interior of the Silver Moon the moment before the
barge began to swing around ; but this was all they
knew about the sloop, except what they had heard
the coxswain say.

"Stand by, all, to lay on your oars !" called Dick,
as coolly as though nothing was the matter with the
Silver Moon, and her passengers were in a frolic
rather than in mortal peril. "Oars !" And every
blade was poised and feathered on a level above
the water.

" Bowman, stand by with the boat-hook ! " continued Dick. " One stroke ! Give way ! "

This single stroke brought the bow up near
enough to the sloop to enable the bowman to fasten
the boat-hook to the gunwale of the helpless craft.
The crew lay upon their oars, ready to obey the next
order, but not one of them manifested the slightest
interest in the Silver Moon, so far as any look or
movement was concerned. Paul Bristol was excited and uneasy, and once he was on the point of
standing up to get a better view of the interior of ·
the sloop. But he remembered the order of the
coxswain in season to restrain himself.

" On board the Silver Moon ! " shouted the coxswain, but not louder than was necessary to make

the skipper hear him above the noise of the wind and the water. " What is the matter?"

" I miss-stayed in going about, and shipped a sea. The boat is half full of water, and I can't do anything with her," replied Bissell, in tones which indicated that he was in utter despair. " The girls are frightened out of their wits, and the water comes in faster than I can get it out."

" Do you want assistance?" asked the coxswain.

" Of course I do !" exclaimed the skipper. " We shall all go to the bottom in a few minutes, for there is a good deal of ballast in the boat."

" All right! We will stand by you," replied Dick.

" Can't you do something more than that? " demanded Bissell, in shaky tones.

" I will send two hands on board to assist you," added Dick. "Dory, you will go on board of the Silver Moon. Take any one you please with you, and report what you think should be done."

" As Paul Bristol is of the least use in the barge, I will take him," replied Dory, as he unshipped his oar.

" He is not the best boatman on board," added the coxswain.

" If he will only obey orders, that is all I want

of him," answered Dory, as he made his way to
the bow of the barge.

A standing order to all hands was never to
stand up in a boat when it could possibly be avoid-
ed, and Dory crawled on all fours, from thwart to
thwart, between the oarsmen.   He was followed
by Paul, in the same safe though undignified manner, for he thought it was not derogatory to follow
the example of the skipper of the Goldwing. The
bowman hauled the boat up so that the two hands
could get on board of her.

Just at that moment all the girls screamed, or,
as Paul expressed it, " squealed," and the lee side
of the Silver Moon rolled under, taking in a barrel or two of water.

" We shall all be drowned ! " shrieked one of
the ladies, as they all sprang out of their seats and
rushed over to the weather side, throwing the boat
out of trim so that she took ·in another barrel of
water over the port side.

" She won't sink yet, ladies, unless you sink
her," said Dory, rather sharply.   " Three of you
on each side, and don't move for your lives. You
will certainly swamp the boat if you don't keep
still.   Don't one of you move again without
orders."

"That's what's the matter," said the skipper. "I can't keep them still."

"They must keep still," added Dory with emphasis.

Just at that moment came another roll, and Dory told Paul to stand in the middle of the boat, and allow none of the passengers to move. He took position near him, and together they kept the ladies quiet, and very little water was taken aboard.

"I am about used up," said Bissell, who was still baling with all his might, though he was nearly exhausted. "I have been throwing out the water for more than an hour."

"You might as well try to bail out Lake Champlain as this boat, while she lies in her present position. The water comes in faster than you can throw it out," said Dory. "Here come our other barges. Don't be alarmed, ladies. Even if the boat sinks, we can save every one of you. Do as you are told, and you shall be made comfortable in a few minutes."

Oscar Chester put the Gildrock about with consummate skill, though the barge shipped some water during the manœuvre. Thad Glovering did quite as well in the Winooski. In a few moments, the

three barges had brought their bows up to the
water-logged sloop. The oars were trailed, and
bunters put over the sides to prevent the boats
from grinding against each other. The Silver
Moon smoothed the water for them a little, and
they rode very easily on the swell.

Dick Short reported to the acting commodore
what he had done, and Oscar said he should not
interfere. At this time, Dory reported the condition of the sloop, and advised that two of the six
ladies be taken into each barge. With great difficulty, on account of the uneasy motion of the
boats, the passengers were transferred to the stern-
sheets of the barges. They were all wet through,
but the commodore would not allow the boats to
leave the scene of the disaster until the safety of
the Silver Moon was assured.

CHAPTER XIV.

THE WORK OF AN INCOMPETENT SKIPPER.

THE removal of the six ladies made the Silver Moon a little more buoyant; but she was in hardly less peril than before, for she rolled even worse than when her passengers were on board. Dory formed a very unfavorable opinion of the seamanship of Bissell almost as soon as he had put his foot over the gunwale of the sloop. Nothing at all was the matter with the Silver Moon. Her rudder was not disabled, and the halyards on the mainsail were in working condition. The craft had lost nothing, but the skipper had lost his head.

But the Beech Hiller did not utter a word of rebuke, or even a critical comment on the management of the sloop. He saw how it was, and understood the situation perfectly, but he did not feel called upon to censure the action which had been taken, or the steps which had been neglected. The craft was in the trough of the sea, and half full of water. He looked about him, and discovered a

145

pair of oars stowed away under the seats in the large room.

"As you are tired out, Mr. Bissell, Paul will bale for a while till you get rested," said Dory, as soon as the passengers had been transferred to the barges, and without any of the delay indicated by the printed page.

"I can do that first-rate," replied Paul, who had wondered what he could do in a sailboat, in the management of which he was comparatively ignorant.

Paul took the bucket; and it is safe to say that he threw out more water than the skipper could have done in his exhausted condition. But the baling appeared to have little or no effect on the large quantity of water in the standing room.

"I am about tuckered out, but what do you think I had better do?" asked the skipper, as he gave up the bucket to the fresh hand.

"I think you had better rest yourself," replied Dory, as he drew out the oars from under the seats. "I will look out for the boat, and I think she will come out of the scrape all right."

."What are you going to do? Do you mean to row her over to Westport?" asked Bissell, panting with his last exertions.

"No; I don't think we should make much trying to row this boat nearly four miles against a head sea," answered Dory.

"But you can't land over on this side of the lake. Just see the waves breaking on the shore in Button Bay. The Silver Moon would be smashed into a thousand pieces," protested the skipper.

"Of course we can't make a landing on a lee shore in this weather," answered Dory, as he went forward.

Bissell could not make anything of Dory, and he looked at him only to wonder what he was going to do. The skipper had always believed that he knew all about sailing a boat; and in a moderate, or even a fresh breeze, he could do very well when everything went along smoothly. But he had never been trained, as the students at Beech Hill had, for seasons of emergency.

The Silver Moon miss-stayed probably because she had not a "good full," had fallen off into the trough of the sea, and rolled herself half full of water before the skipper thought of doing anything to overcome the difficulty. Under these trying circumstances, he was not instructed either by precept or actual trial what to do.

The throat halyard of the sail, Dory found when

he went forward, had not been cast off. He got hold of the peak halyard and hauled on it till he got a good set on the mainsail.

"What are you about?" cried Bissell. "You will upset her as sure as you live! I let that part of the sail down because it blowed so hard. I was going to lower the whole sail, but I had n't time. I was afraid the boat would sink if I did n't bale with all my might."

Dory secured the peak halyard without paying any attention to the shout of the skipper. When he had put on the half hitch, he went aft again.

"Let the sail down just as quick as you can!" yelled Bissell.

"I think not," replied Dory quietly.

"If you don't, I shall lower it myself. I can't stand this!" added the frightened skipper.

"This boat belongs to you, Mr. Bissell, and you can do what you please with her, and I shall not quarrel with you about it. If you don't want any assistance, Paul and I will return to the barge."

"But I don't want anybody to sink her," protested the skipper.

"I shall not sink her; but it is all nonsense to think of baling her out while she is in the trough of the sea, shipping water as fast or faster than you can throw it out. I am not willing that Paul should wear himself out for nothing."

" That 's the only way to keep her from sinking."

" You have a perfect right to your own opinion, as I have to mine. I am confident that I can get the sloop out of this scrape, if you will allow me to do it; if not, Paul and I will return to the barges."

" But I want you to help bale out the boat. She will sink if the water rises any higher in her," the skipper insisted.

At that moment, a heavier wave than usual struck her, and rolled the boat over till she took in at least a barrel of water. Bissell sprang from his seat as though he expected the Silver Moon to go to the bottom at that instant. But she did not sink, though her floating power seemed to be very nearly exhausted.

" Do you see that?" demanded the skipper, as though he regarded the water just shipped as a triumphant vindication of his opinion.

" I see it; and I should say she will go down after she has done that thing once or twice more," replied Dory calmly. " Say quick whether I am to get the sloop under way, or whether I am to leave her! Do as you please about it. The barges will save you when she sinks."

" I will see what you are going to do," replied Bissell doggedly ; and he was by no means convinced, though he was satisfied that he could do nothing alone to save the craft.

Dory made no reply, for he was rather disgusted with the obstinacy and nautical ignorance of the skipper. Taking one of the oars in his hand, he went to the side of the boat and hailed the commodore of the barge-fleet.

" I am going to get under way now, for there is nothing the matter with the boat," said Dory.

" All right, Dory," replied Oscar, as he looked about him to determine the best way to separate the barges, and get them clear of the Silver Moon.

They were all lying close together, the Gildrock being nearest to the bow of the sloop. The commodore ordered the Winooski to cast off first. With her port oars she kept her head up to the sea till those on the other side of the boat were clear of the Marian, which was next to her. When her twelve oars were in the water, the boys backed her clear of the other boats, and then the crew were ordered to lay on their oars. In the same manner the other two boats backed into safe positions. The barges pitched tremendously, and the ladies were inclined to " squeal." When any of them attempted to stand up, the gallant coxswains pulled them back into their seats.

"Are they going to quit us?" asked Bissell, as he looked with something like dismay upon the movements of the barges.

"We could not get the sloop under way with all those boats hanging to her," replied Dory. "They will be here quick enough if they are wanted. But we are not going to ship any more water; we shall leave the rest of it in the lake."

"I don't see how you are going to help taking in all the water that rolls in," growled the skipper.

"If you keep your eyes wide open tight for one minute more, I will show you that the Silver Moon is the mistress of the situation, and Lake Champlain will have to stay out in the cold," replied Dory, as he shipped his oar between a couple of thole-pins on the quarter.

"There is another oar under the seats," added Bissell.

"One is enough to do the business," said Dory, as he began to pull with all his might.

The water-logged craft moved as though it had been fastened to the bottom. The boom was still hanging out at right angles with the keel, and the main sheet was flopping about in the water.

"Can't I help you?" asked Bissell.

"You may take the other oar, if you please."

The skipper pulled out the oars, and was about to ship it on the weather side when Dory interposed, and, shifting his own blade to the fore rigging, directed his companion to ship his between the thole-pins he had left.

"On this side? Who ever heard of rowing in that way?" blustered Bissell. "Both oars on the same side!"

"We have no time to argue the question now, if you want to keep your boat on the top of the water instead of the bottom," said Dory sharply.

Bissell did as he was told, though he had no faith in rowing with two oars on one side of the boat. The united strength of the two was immediately seen in the motion of the boat. Her bow was soon thrown up to the wind, and then the boom swung in over the standing room. This was the acting skipper's object, and as soon as he could reach the main sheet, he dropped his oar. Grasping the rope, he carried it aft, and got a turn with it over the cleat above the rudder head.

The sail filled as soon as he hauled in the sheet, and Dory got hold of the tiller. The sloop heeled over till Bissell declared that the new skipper would upset her. Paul continued to bale with all his might. Dory trimmed the sail down as flat as he could, and soon had perfect control of the craft, heavy as her movements were in her water-logged condition. He was obliged to touch her up in the fierce blasts which struck her, but he had her well balanced, and she did not realize any of the evil predictions of her incompetent skipper. On the other hand, she did not dip up any more water over her gunwale, and all that came into her was in the form of spray.

At Dory's suggestion, Bissell got out a firkin in which the lunch for the party had been brought on board, and assisted Paul in the work of baling. But there were hogsheads of water in her, and the process of relieving her was very slow. The three barges were still laying on their oars, watching the movements of the Silver Moon. Dory ran for Button Island, which was separated from the main land only by a narrow channel. Slacking off the sheet, he ran her aground in the sand behind the point, where the water was as smooth as the lake in a calm.

"She is aground!" exclaimed Bissell, as the keel grated on the bottom.

"I know it; and that is where I intended to have her," replied Dory, as he left the tiller.

Bissell began to protest that he would not have the boat aground; but the new skipper paid no attention to him. Taking the bucket from Paul, he told him to rest. Dory worked hard at baling for half an hour, and Bissell did his part as soon as he found that his protest was not heeded. By this time the water had dropped so it had to be dipped out of the well. It was all out at last, and the well sponged dry. To the stupid astonishment of Bissell, the Silver Moon was again afloat.

CHAPTER XV

FOR about three-quarters of an hour the three barges had waited, pulling just enough to keep their heads up to the sea. By this time, the ladies had become accustomed to the motion of the boats; and, in spite of their wet and bedraggled condition, they began to be very jolly. The long boats rose and fell with the waves, and occasionally the spray was dashed over the bows, and carried the whole length of the craft. At last, they began to sing, and the students joined them. It was getting to be a very merry time on the rough waters, but the confidence of the crews inspired the ladies with courage.

As soon as the Silver Moon was free from her burden of water, Dory examined the sail, and, finding it all right, he shoved off. As Paul Bristol had said, the Silver Moon was about the size of the Goldwing, and was a very good boat.

"I have not the least idea that you can sail this

155

boat over to Westport in this blow," said Bissell, as Dory shoved the sloop out of the shoal water.

"What's to prevent?" asked the new skipper coldly.

"It blows too hard, and the sea is too heavy for any boat, I don't care how good she is," answered the owner of the Silver Moon. "I think this boat is as good as any of them, but I had rather walk twenty miles than cross Lake Champlain in her in this blow."

"The sloop belongs to you, sir, and you can do as you please about crossing," answered Dory. "You can anchor and stay here till to-morrow if you like."

"Do you think it is safe to cross the lake in a sailboat when the wind blows as it does now, and has since ten o'clock?" asked Bissell.

"I have been off Burlington, where the lake is twelve miles wide, when the sea was a good deal worse than it is here, and I did not think I was in any greater danger than if I had been on shore."

"If you are not afraid, I ought not to be," added the owner of the craft, evidently laboring to stimulate his courage. "I guess I will risk it, as those barges will be near enough to pick us up if anything happens."

"All right," answered Dory, as he hauled down the sheet, and let the boat go ahead. "Why did you come out in such a blow, if you don't think it is safe?"

"The girls had n't anything to do to-day, and wanted to go down to Port Henry. We started at six o'clock this morning, and got there at eight. It began to blow pretty hard by nine, and we started back, though we intended to stay at the Port all day. I got along very well, though the girls were scared, till we got down to Barber's Point; and after that it blew like tophet."

"You had the full rake of the wind across North West Bay then," added Dory.

"Yes; and I had to beat all the way home right against it," continued the skipper. "I thought we should tip over every minute. If I let her off enough to make her go ahead, she tipped so that the girls all screamed. When I was half way over to Button Island I tacked, but the boat would not come about. She lay there with her sail banging. Then the wind caught the sail again, and tipped her so she took in some hogs-heads of water. She got in between the waves, and began to roll like a chip. I thought it was time to haul down the sail, and I went forward

to do so. Then the main sheet run out, and I
could n't get it again."

" It was a bad situation," added Dory, as Bissell
paused and looked at him, apparently as if to ascertain what he thought of the skipper's management;
but Dory expressed no opinion.

" I let go the peak halyard, and did the best I
could to get the boom in, but I could n't do a thing
with it. The boat kept rolling in the water all
the time, and I had to take the bucket and bale
with all my might. I was afraid to haul the sail
down then, for it would have gone into the water,
and helped drag her over on one side."

Dory, as an expert, had a very decided opinion
in regard to the skipper's management; but he did
not feel called upon to express it, for Bissell was an
obstinate man, and he did not care to dispute with
him. The Silver Moon was running out close-
hauled from the lee of Button Island, which carried her to windward of the fleet of barges. Dory
had taken the helm when she got under way; and as
long as the skipper did not object, he retained it.

" The boat don't work very well without the
jib, and that was what made all the trouble," continued the skipper. " But it blowed so like all
possessed, that I could n't carry it."

Dory doubted whether this was all, or even the principal trouble, but he made no remark. He was not satisfied with the working of the boat, and without saying anything to the skipper, he put her about, and ran back to the lee of the island. Getting her forefoot on the sand far enough to hold her, he let go the halyards, and lowered the mainsail a few feet.

"What are you going to do now?" asked Bissell, who had watched the movements of the acting skipper with interest and anxiety.

"I am going to put a single reef in the mainsail. Where do you keep the reef pendant?" replied Dory.

The skipper had no reef pendant, and probably did not know what it was. But Dory found a couple of ropes which answered his purpose. Having lashed down the clew and tack, with the assistance of Paul, he tied the reef-points. Hoisting the sail up to a good set, he shoved off with the boat-hook, and was soon standing out from the shore again. Keeping his place at the helm, Dory stood out to the fleet of barges. The wind had not abated a particle of its force, and even with the reef in the mainsail, she was inclined to bury herself in the waves. Dory was not yet satisfied, and under his direction Paul pulled up a couple of loose boards in the floor of the standing-room, and lifted out a couple of cast-iron blocks of ballast. These were placed in the stern, and the bow was lifted a little more out of the water.

"What's all that for?" asked Bissell.

"To change her trim a little," replied Dory. "She was ballasted too much by the head. She works better now."

The sloop was less inclined than before to bury her bow in the waves, and was more buoyant forward. She dashed ahead at a gallant speed, and in a few minutes she was approaching the barges. The passengers in the stern sheets of the boats were very merry by this time, and seemed to be actually enjoying the motion of the boats. As the sloop came within hail of the barges, the ladies struck up "Roll on, silver moon," in which all the students who could sing joined, and it made a very effective chorus.

"Very appropriate," said Dory, laughing; "but she don't do that now. She has done rolling for the present."

"The girls are jolly enough now; and they don't seem to be a bit scared," added Bissell.

"Will you take them on board again?" asked the acting skipper.

"I don't know; what do you think?"

"If you don't know, you had better leave it to them," replied Dory. "Hail them, and ask them what they will do."

"I say, girls, are you ready to come on board of the Silver Moon?" called Bissell, in a loud voice, as the sloop passed astern of them.

"No!" screamed the whole of them, almost with one voice. "We are going back to Westport in the barges," added one of them.

"Just as you like," returned Bissell.

"Roll on, silver moon," the girls struck up again, and the boys took up the chorus with enthusiasm.

"That settles it," added Bissell.

"Then we may as well return to the Marian, Paul," said Dory.

"I am ready to do just as you say," replied the spare hand.

"You don't mean to leave me, do you?" demanded Bissell, aghast at the proposition. "I don't want you to go."

"I thought you might prefer to handle your own boat," suggested Dory.

"I guess I'd rather have you take her over to Westport, if you will," replied the skipper, anxiously.

"Even if we remain on board of the sloop, you had better take the helm," added Dory, who had some curiosity to see how Bissell worked the boat.

"I'd rather have you steer her. I don't know but you can manage her better than I can."

"You ought to be able to handle your own boat better than any one else can. You have sailed her more than any other person, and a boat is something like a horse, and does better in the hands of one who is used to her."

"I bought the Silver Moon last year, and got a man to show me how to manage her. I was out in her every day last summer, but I never went out when it blew very hard. Folks say it is dangerous sailing on Lake Champlain, there are so many currents and flaws from the hills."

"There is no doubt about the flaws and currents, but I look upon them as bugbears. A skipper must keep his craft in hand all the time, and then he is ready for flaws and squalls."

"One of the girls has taken your place at the stroke oar, Dory," said Paul, who was watching

the barges as they began to move over the savage
waves.

"So I see," replied Dory. "Dick is coaching
her, and I have no doubt she will make good my
absence."

"It is Susy Wellington; and she knows how
to row better than most of the men," added Paul.

At this moment, the acting skipper went in
stays, and though he had given her a good full,
he had some doubts about her; but she came up
to the wind handsomely, and went on the port
tack as promptly as the Goldwing could have
done it. As she filled away, she heeled over
till her washboard was almost buried; but she
righted a little in a moment, and dashed off on
her course like a racehorse. She rose and fell
on the waves, with her gunwale under all the
time, but with eight inches of washboard above
the water.

The wind was rather flawy, and, of course, the
boat heeled over more when the puffs struck her,
so that most of the washboard was sometimes
under water. But the sloop, in her altered trim,
was as steady as an old horse on a smooth road.
As the Silver Moon was close-hauled, she struck the
seas constantly; and the waves broke with no little

noise against her bows, tossing the spray from stem to stern.

Bissell watched the lee washboard all the time, and seemed to be very nervous. He did not heed the singing in the barges, which greatly interested the acting skipper. The owner evidently expected the sloop would ship a sea every moment, which would fill her half full of water as she had been before. But she tore along on her course without taking in a drop of water over her lee side, unless when a wave broke there, and spit the spray over the washboard.

"You will put the Silver Moon on the bottom before you get her over to Westport," said Bissell, when he could hold in no longer, and his mental excitement had become intense.

"I certainly shall not do it after we get to Westport," replied Dory, with a smile. "But if you wish to take the helm, of course you can do so. I suppose you can swim, Paul?"

"I could swim in Sandy Bay, but I don't think I could in these waves," answered Paul.

"I don't think it is safe to sail along in this way with the gunwale under water all the time," said Bissell, as the water rose nearly to the top of the washboard.

"Then you take the tiller, Mr. Bissell," replied Dory, rising and offering his place on the weather-side to the skipper.

Bissell was clearly full of doubts, but he took the tiller. His first movement was to put the helm down a little, so that the reefed mainsail began to shake slightly, and of course the gunwale was lifted out of the water. He kept the tiller shaking all the time, as the boat was in danger of broaching to.

When he had steered about a quarter of an hour, it was seen that the barges were rapidly overhauling the Silver Moon, though she gained on them while Dory had the tiller. But it was necessary to tack, and the skipper put the helm hard down. The sail shook, and the boat did precisely what Dory knew she would do — she miss-stayed, and then began to roll in the trough of the sea. She had little headway when the helm was put down, and her momentum was not sufficient to carry her around against the head sea.

The water began to roll into her on the sides; but Dory seized one of the oars, and with a few smart pulls, threw her head up into the wind. The instant the sail began to fill, which it did with a rush, Bissell put his helm hard down. Dory plied the oar once more.

" I wish you would take the helm again," said Bissell.

"I will, if you say so," replied Dory, who had entirely satisfied his curiosity in regard to the seamanship of the skipper.

CHAPTER XVI.

DORY DORNWOOD GIVES A LESSON IN BOAT-SAILING.

IT was absolutely certain that Bissell was not a reckless and over-daring skipper, which is often the most glaring fault of those in charge of sailboats. He erred in the opposite extreme, — he was too timid. He had not pluck enough when it was blowing fresh to keep his sail full. The barges were overhauling her, because she had hardly any headway ; and when she went in stays, she had not speed enough to meet the fierce waves.

Dory took the helm, while the skipper, with the bucket, dipper and sponge, soon removed the water from the well. The sail was permitted to fill, and the Silver Moon dashed on her course at a lively rate again. The barges pulling against a head sea could not keep within hail of her when she was on her long tack.

"I have about made up my mind that you can handle this sloop better than I can," said Bissell, when he had wiped out the well with the sponge.

Paul Bristol burst out into a loud laugh at this remark.

" What are you laughing at, Paul?" asked the skipper, looking rather severely at the spare hand.

" I was only thinking it had taken you a long time to make up your mind, Mr. Bissell," replied Paul, suppressing his risibles when he saw that Dory looked as serious as a judge.

"I suppose you think I am not much of a skipper, Paul," added Bissell, evidently annoyed by the laugh of the spare hand.

"I don't know much about sailing a boat, and I think I had better not say anything," answered Paul prudently.

" I never got into such a scrape before with the boat," continued the owner. "This boat will be for sale after I get ashore."

" She is a very good boat, and works well," said Dory ; but, cautious in regard to offending the skipper by any criticisms, he was not willing to have the boat blamed for the fault of the man.

" I always thought so myself till to-day ; and I have always believed I could handle her better than any other man. It goes a little hard with me to give in to one boy and have another laugh at me," replied the skipper. "I should like to have you tell me just what ails my management of the boat."

" I don't volunteer any criticism ; but if you will not be offended with a boy for expressing his mind plainly, I will do so," added Dory.

" That's what I want you to do, and I shall not get mad, though it hurts for me to give in on handling the Silver Moon.

" You are just a little too careful; and that is what has made the mischief every time," Dory began. " If you don't give your boat a good full, she won't go about in stays. That was the trouble when you had the ladies on board."

" The man that showed me how to sail a boat said I could not be too careful," protested Bissell, astonished at the remark of the acting skipper.

" I don't quite agree with him, though he is right in the main. Most of the accidents happen because the skippers are careless. Your sloop was a little out of trim. When it blows too hard for you to carry the jib, you must put a single reef in the mainsail. With a whole mainsail, the mast ought to be farther forward. Since I shifted the ballast, she carries a stronger weather helm."

It was necessary to tack again, and Dory explained more fully what he meant by a " good full," and then put the helm down. The sloop's head flew up into the wind at a lively pace, and the mainsail went over; but the helmsman righted the helm, and met her with it so that she should not fall off too far, thus putting her lee gunwale under.

Bissell was deeply interested, and began to learn what he had not before acquired. He took her, and made the next tack, under the direction of the acting skipper, himself. From that time he retained the helm, and Dory continued to instruct him until the Silver Moon got into comparatively still water.

Dory and his pupil had made more tacks than were necessary, in order to illustrate the subject. The boats came up with the sloop just as she was going in at the wharf. The ladies and the crews were still singing, and their merry voices attracted quite a collection of people.

" When I was over off Button Island I did not expect ever to see Westport again," said Bissell, after he had made the Silver Moon fast at the wharf. "I have only ten dollars in my pocket, now, but I want you to take that and let me owe you another ten."

" You must excuse me, Mr. Bissell," protested Dory.

"It ain't enough, I know; and I will make it up to fifty dollars when I get to the store," added the skipper.

"It is enough, and more than enough, Mr. Bissell. I cannot take a single cent for that kind of service. Captain Gildrock would put me into the brig if I took money for assisting anyone in distress on the water," said Dory earnestly. "Sailors are bound to help each other always when in danger."

The store-keeper pressed Dory quite warmly to take the money, but the latter was as firm as a rock. Then he tried to give a few dollars to Paul, but the spare hand, though he wanted the money for his mother, took his cue from Dory, and refused to take a cent. Bissell expressed his gratitude in very warm terms, and said he should like to take some more lessons in sailing a boat from such a skilful master. He would gladly pay for the time and trouble, and he concluded not to sell the Silver Moon at present.

It was now nearly two o'clock, and the students thought it was about time to attend to the contents of the lunch baskets. The ladies had been landed, and were profuse in their expressions of delight at their trip in the barges. The Beech

Hillers landed, and camped under a tree to dispose
of their lunch. Dory and Paul joined them, and
it took some time for the former to explain what he
had done on board of the sloop. He did not say
any unpleasant things about the skipper, or dwell
upon his mistakes.

Before the boys had made any deep inroads into
their stock of provisions, Bissell and the ladies appeared laden with ice cream, pies, and cake, which
were a welcome addition to the lunch. The " girls,"
as Bissell called them, waited upon their deliverers,
and gushed over the delightful time they had had
in the barges.

" There come the Chesterfield barges," said John
Brattle when the clock indicated half past two.
" I wonder what they are going to do over here."

" I thought they had business about this time
over at Sandy Point," added Tuck Prince.

" Don't say a word about the cottage, fellows,"
interposed Paul Bristol, with a good deal of earnestness. " I know what they come here for; at
least, I think I know."

" Why don't you let on then, Paul?" demanded
Phil Gawner.

" They have come over here after Major Billcord,
for I am sure he will want to see the cottage pitched

into the lake. I am almost sure now that none of them know the house is gone," replied Paul, rubbing his hands with delight when he thought of the disappointment of his oppressors.

"If you like, Mr. Bissell, I will sail down the lake with you as far as Sandy Point," Dory proposed, while the boys were digesting what Paul had said. "The coxswain consents to my absence; but I must return to Beech Hill in the Marian."

"All right, for I want very much to see you sail the Silver Moon with a heavy wind on the beam or over the quarter," replied the store-keeper. "But I must go up to the house and change my clothes, for I am as wet as a drowned rat."

Bissell hastened to his house, which was only a short distance from the head of the landing. The Chesterfield barges had just reached the wharf, and the young gentlemen were coming up the steps. The boats had pulled around under the lee of the land, so that they had not been seen until near the wharf.

The Chesterfield students formed a procession on the wharf, and it was evident that they intended to escort Major Billcord, who was fond of parades, to the boats. The ladies waiting on the students from the other side said this was the meaning of

the procession, which was not a strange sight in the
streets of the town.

"You will take no notice of them whatever, fellows," said Commodore Chester very impressively.
"If they salute us properly, which they are not
likely to do, we must be as polite as they are, and
more so, if possible. If they call us 'tinkers' and
'chip-makers,' which they are more likely to do,
make no answer of any kind. I will report any
student who utters an offensive word to them.
You all know that this is the order of the principal, and not mine."

For some reason the procession of Chesterfields
did not pass near the tree under which the Beech
Hillers were lunching. Paul pointed out the elegant mansion of Major Billcord, and the students
of the institute marched in that direction.

"While we are waiting for Mr. Bissell, I should
like to go up to his house and get my sister's valise, which she left there," said Paul to the coxswain of the Marian.

Dick consented, though he would not have permitted any of the crew to leave without a good
reason for it while the Chesterfields were so near.
Paul hastened up to the main street. He saw the
institute students halt in the grounds of Major

Billcord's mansion. They broke ranks, and the magnate was talking to them.

"I will give twenty-five dollars to any student or party of students that will capture that young scoundrel, Paul Bristol, and hand him over to me at Sandy Point." This was what the great man said to a group of half a dozen of the students.

The party in front of him promised to carry out his wishes if he would not mention the matter to the rest of the students.

CHAPTER XVII.

PAUL BRISTOL saw the conference between the six Chesterfield students and Major Billcord; but he could not hear what passed between them, and had no suspicion that he was the subject of remark. None of them saw Paul when he entered the gate to Mr. Bissell's premises, though his curiosity prompted him to stand there a few minutes to observe the proceedings of the party.

He expected to see nothing more than a sort of reception of the magnate, who was evidently to be their passenger to Sandy Point, and the students proposed to take him to the scene of the afternoon's sport in state. But Major Billcord appeared to have selected the six ruffians best suited to the undertaking in which they were to engage. He had called them aside, and made his offer to them.

Those who were near enough to the magnate to see his face could not help noticing that he had a

176

pair of black eyes. In this respect he was the counterpart of his hopeful son, though the mourning of the latter was of a deeper shade than that of his father. The major had remained in his elegant mansion all the forenoon, for he was more modest in the display of the weeds under his eyes than he was of his person generally. Doubtless he had often looked in his lofty mirrors to observe the condition of his face.

He did not like the looks of himself with the marks Paul had left on his face, for they certainly added nothing to the dignity of his expression. He was a pompous, overbearing, and tyrannical man, and every time he saw his mourning organs they filled him with wrath, and inspired him to seek a wholesale revenge. He did not give a thought to the insult his son had offered to Miss Lily. She was of not the slightest consequence, and it would have been quite proper, in his opinion, for her to submit in silence to the pleasure of the reckless young man.

On the way home from the scene of his signal defeat at the hands of Paul Bristol, he had called at the institute, and intimated that he had a mission for the students, at the point, on the following day. He wished them to call for him at his house

in the afternoon, and he would insure them an
hour or more of the liveliest recreation. He did
not say what he had in view, and he had cautioned
Walk not to mention the business in which the
young men were to be engaged.

As Walk went home with his father, he had no
opportunity to let the cat out of the bag, even if
he had been so disposed. Father and son had
spent the rest of the day in studying out an ade-
quate punishment for Paul. If they could have
" hung, drawn, and quartered " him, it might have
satisfied them. Walk suggested that he should
be prosecuted, and that the justice would send him
to prison for a month or two. But his father saw
that such a course would bring out the whole story
of the son's assault upon Miss Lily, and the judge
might not regard the affair in the same light as the
sufferers did.

They could agree upon nothing, but before
morning Major Billcord had devised the scheme
he had now taken the first step to carry out. He
thought it wise not to implicate his son in the outrage, for he might be prosecuted and compelled to
pay a fine for himself and those he employed to do
the actual work. The vengeance of the magnate
was to be administered to Paul at Sandy Point.

The plant for the black eyes had been set out near the doomed cottage, and it was proper that the punishment should be inflicted on the same spot.

Walk Billcord had not been in condition to return to the institute that day, for he did not care to put his mourning on exhibition, and to answer all the questions that it would call forth. But he was going to Sandy Point in the boat to which he belonged, for he was anxious to take part in the destruction of the cottage. After the students marched up to the mansion of the magnate, they had informed him of the presence of the Beech Hillers, and of Paul, in Westport. The story of the Silver Moon's mishap had been related to them at the wharf when they landed. The students from the other side had come to the town to convey the ladies from the disabled boat; and this sufficiently accounted for their presence.

Nothing was said about Paul, except that he had come in the sloop. Major Billcord had no doubt that his stalwart foe was still staying at the point, and he had arranged his plan on the supposition that he would be found in that vicinity. But when he was informed of his presence in Westport, he had been obliged to make a slight change in his scheme. He had not intended to mention it till the students landed at Sandy Point. He had before selected his ruffians, and he was simply obliged to make his offer a little sooner than before arranged.

Paul Bristol went into Mr. Bissell's house and obtained his sister's valise. As he was about to depart, the owner of the Silver Moon came down stairs in his changed dress, with a letter in his hand.

"Paul, do you know where Captain Bleeker lives?" asked Bissell, as soon as he saw the boy with the valise in his hand.

"Of course I do," replied Paul. "I used to work for him on his place when he had anything for me to do."

"I am going to Sandy Point in the sloop, and Dory is waiting for me," continued the skipper. "I am in a hurry to be off, and if you will go round by Captain Bleeker's and leave this letter at his house, I will carry your valise down to the boat. I will put it on board of the Marian."

"All right, if you will tell the coxswain to wait for me; for I suppose I am to go in the barge," replied Paul, as he took the letter.

Bissell hastened to the wharf with the valise, and Paul started for his destination, which was on a street in the rear of Major Billcord's mansion. The

procession had re-formed in the spacious grounds
to escort the magnate to the wharf. The six ruffians had been excused from marching in the line,
by request of the major, and they were consulting
in regard to their mission in the street in front of
the house.

Paul was obliged to take a cross street to reach
the house of Captain Bleeker, and he had to pass
within a few rods of the elegant mansion. As he
turned the corner, the chief of the six ruffians, who
was called Buck Lamb by his fellow-students, discovered him, and the conference came to a sudden
conclusion. The time for council had passed, and
the time for action had come. Buck Lamb was an
acknowledged leader, and, without any appoint-
ment as such, he assumed the position and began
to give off his orders.

Paul was in sight, going up the cross street with
a letter in his hand. The back street ran parallel
to the main street, and the object of the attack
must be going to some house in that direction.
Buck sent two of his force to the cross street next
beyond that taken by Paul, with orders to inter-
cept the victim if he went that way. Two more
were to remain near the mansion of the magnate,
and Buck himself, with Ham Jackson, followed

Paul. One of the two parties was sure to meet him, or if they failed, by any accident, the pair on the main street were in position to capture him. It was a quiet little place, and there was scarcely a person in the streets after the procession had marched to the wharf.

Paul Bristol, all unconscious of what had been done to make him a prisoner, walked with a rapid step towards the house of the person to whom the letter was addressed. He had not noticed the movements of the six ruffians, or even that any of the Chesterfields had been left behind. He was thinking that the students would soon reach Sandy Point with Major Billcord, and he was engaged in picturing their astonishment and disappointment when they discovered that the cottage had taken to itself wings, and that the locality had assumed its original appearance.

He went up to the door of the house, and rang the bell. It was answered by Captain Bleeker himself. He seemed to be somewhat surprised when he saw Paul, for he had been expecting another person.

" Is that you, Paul? I was in hopes that it was Bissell; for I expected an important letter as soon as the mail got in, and he promised to bring it over to me," said the captain.

"He asked me to bring the letter over, and here it is," replied Paul, as he handed the important missive to him.

"Good! It contains a draft which I need as much as I need the air I breathe," added Captain Bleeker, as he took the letter, and thrust his hand deep into one of his trousers pockets, drawing forth a quarter. "I am more glad to see you than I should be to meet my grandmother, who died twenty-five years ago. Here is something to prove it;" and he handed him the quarter.

"I don't want anything for this errand; I only did it because Mr. Bissell asked me to."

"Take the money," said the captain imperatively. "If you come over here in about a week, I shall have something for you to do, for it will be time then to hoe the garden."

"I don't think I can come, sir, for I have a place now, with steady work, on the other side of the lake," replied Paul.

"All right," added Captain Bleeker, as he broke the seal of the letter, and proceeded to close the door, manifesting no interest in the messenger's new position.

Paul put the quarter in his pocket, thinking there had been no time in two years when it was so little needed as at present, thanks to Captain Gildrock. But he did not lose a moment, for he thought that by this time the crew of the Marian might be waiting for him. He walked at his most rapid pace up the street in the direction by which he had come. There was not a person to be seen in the back street, though Buck Lamb and Ham Jackson had reached the corner.

Paul saw them approaching him on the same side of the street. If he had seen them in the neighborhood of the institute, he might have known them; as it was, he did not recognize them, though they wore the barge uniform. Being in a hurry, he deserted the sidewalk to cut off the angle at the corner of the street. But the two ruffians promptly placed themselves in front of him in the middle of the highway.

"Stop where you are!" said Buck Lamb, in an imperative tone.

"What am I to stop for?" asked Paul, with a smile, and with the simplicity of an infant.

"You are to stop because I order you to do so," replied Buck, who was of the genus bully, and could not well help manifesting authority, whether he had it or not.

"As I am in a hurry to join my boat, I don't

think that is a sufficient reason for my stopping," replied Paul, with abundant cheerfulness. "If you will excuse me, I had rather not stop just now."

"But I order you to stop!" said Buck savagely.

"Oh, you do!" added the intended victim. "Then I must take the liberty to disobey your orders."

"When I order you to stop, I mean to enforce my order," said Buck, with his teeth set fast together.

"I can't stop to jaw with you now; for, as I told you, I am in a hurry," replied Paul, beginning to be a little indignant at the interruption.

"If you move another step, I shall hit you," continued the bully, placing himself in front of the victim, with his fists clinched ready to execute his threat.

Paul dodged back, and attempted to pass the ruffians, but Buck got in front of him again.

CHAPTER XVIII.

THE RESULT OF AN UNEQUAL CONFLICT.

BUCK LAMB evidently considered himself as more than a match for Paul Bristol, for neither Walk Billcord nor his father had given the students the particulars of the battle at Sandy Point. All the magnate said was that he and his son had been insulted and assaulted by the ungrateful son of the woman he had harbored on his land. Buck was a stout fellow, who had the reputation of possessing scientific skill in the noble art of pugilism, and who was ready for any fellow of his avoirdupois, either with soft gloves or with hard gloves, or with no gloves at all.

Happily, Paul had no knowledge of the reputation of the bully, which the more knowing of the students would have said was only reputation. In his ignorance of the accredited accomplishments of Buck Lamb, Paul was not in the least afraid of him. Possibly, though not probably, if he had known what a bruiser the leader of the six ruffians claimed

186

to be, he would have taken to his heels and escaped in the best way he could, or expressed his willingness to obey the imperative order of his assailant.

Buck not only intercepted his intended victim, but he levelled a blow with his iron fist, as he called it, at the modest nose of Paul Bristol. This was enough to satisfy the son of toil, who had often hoed corn and potatoes in the garden near the corner, that his opponent meant business. He parried the blow aimed at him successfully, and it was the right time for him to have returned it; but Paul did not yet mean business, for he was not a fighting character, and despised the whole subject of pugilism. Instead of striking, he looked about him for the means of escape, and discovered two more of the ruffians running with all their might towards the scene of the encounter. They wore the uniform of the barges of the institute, as did Buck and Ham. There was no chance of escape, and Paul was beginning to get a glimmering idea of the purpose of the assault. He concluded that the fellows before and behind him were to punish him for what he had done the day before at Sandy Point.

Buck Lamb did not wait for the second pair of the ruffians to join him; and in that he made a bad mistake for himself. His blow had been parried, and he began to think that his antagonist had some skill in the sublime art of pugilism; in this he was again mistaken, for Paul had never learned the first thing about it. He was fully roused, and he aimed a second blow at Paul, which was not as successfully warded off as the first had been, and he received a portion of its force in his right cheek.

Perhaps this hit was needed to render Paul fully equal to the needs of the occasion; at any rate, it roused all the tiger of his nature, and then he was ready for anything that might come. He attempted to parry the blow with his left arm; but as soon as the hit was felt on his cheek, he struck a tremendous blow with his right hand. It was the stroke of a son of toil, whose muscles had been hardened by constant labor. It fell between the eyes of the bully, which seemed to be a favorite locality for Paul.

The pugilist of the institute reeled backwards, and then fell over into the dirt in the middle of the street. Doubtless he saw all the stars his vision could encompass, and his ideas were awfully confused. At any rate, he did not "come to time," but lay on the ground where he had fallen.

Ham Jackson was appalled at the result of the first onslaught, which was accomplished in a tenth part of the time it has taken to narrate it. Perhaps he was not a pugilist; but it was certain from his actions that he was not prepared for the state of things now existing. He seemed to be quite as much bewildered as his companion in the dirt. But he recovered himself more quickly, and rushed upon Paul with the apparent intention of seizing him by the collar of his coat. But the son of toil did not know what he meant, and only saw him coming. Without waiting to inquire into his purpose Paul gave him the counterpart of the blow which had upset Buck Lamb. The effect was precisely the same, and Ham went over backwards.

The two students approaching from the rear were only a few rods distant, and Paul did not wait for them to come any nearer. As he would gladly have done in the first place, he took to his heels and ran as fast as he could towards the main street. Before he reached it, he looked back to see if his enemies were pursuing him. The two who had just come up with their unhorsed companions were picking them up, and there was no effort yet made to overtake him.

Paul slacked his speed to a walk, and as he approached the main street he saw two more students in the barge uniform. The instant they discovered him, they rushed to a position in front of him. One of them commanded him to stop; but he declined to do so. One of them attempted to seize the collar of his coat, and Paul felt obliged to hit again. His blow was parried better than either of his former assailants had been able to do it, but he repeated the attempt with success. The blood spurted from the nose of the foremost student, but he was not upset by the shock.

Both of them came upon him then, the second putting one of his feet between Paul's legs while he was attending to the first, and tripping him up. Down went the son of toil, and the two ruffians pounced upon him with the intention of holding him till the rest of the party joined them. But Paul did not hold still worth a cent; and with his great strength he shook off his assailants.

The last couple were more plucky than the first had proved to be, and they followed him up very closely. The victim rained blows upon them without regard to the belt, hitting them where he could. He was furious, and raging like a lion. It was a very uneven combat, and Paul was rapidly exhausting himself. The second of his present antagonists used his boots almost exclusively. He kicked the son of toil in the shins, and labored to trip him up again. Just as Paul had given the one who used fists a blow between the eyes, which upset him, the leg operator succeeded, by an attack in the rear, in bringing the victim to the ground.

As soon as Paul was down, Mad Twinker, who was the one who demonstrated with his boots, jumped upon him, and clawed his fingers into the throat of the unfortunate son of toil. Just at this moment the party from the back street came up, and the two fresh ruffians assisted Mad in securing the victim. They had obtained a couple of pieces of bedcord at the house of the major, and they tied his arms behind him.

Paul was so exhausted that he could make no further resistance, and he submitted to be bound. His breath was hardly shorter than that of his last opponents, who were now wiping the blood from their faces. The ruffians had earned their money, so far as the capture of the victim was concerned; and it only remained for them to deliver the prisoner to Major Billcord at Sandy Point.

Two of the ruffians went to a pump and wet their handkerchiefs, with which they washed their own faces. Paul's handkerchief was taken from his pocket, and the stains of blood were removed from his battered face. In a few minutes they had wiped away the traces of the conflict. Then two of the students, one on each side, took Paul by the arms and marched him towards the wharf. When they came in sight of it, they saw that all the barges were manned, and had pulled a short distance from the shore, where they awaited the absent members of the crews. The Beech Hillers had taken position off the end of the pier, while the Chesterfields were near the land.

There was still quite a collection of people on the wharf, drawn thither for the purpose of seeing the barges, and possibly expecting a race or a fight, the latter being more probable in the opinion of the spectators; the battles of the previous season having been duly reported in the town. The Silver Moon, with Bissell and Dory in the standing-room, was running with the wind on her beam for Scotch Bonnet. The six ruffians halted when they came in sight of the wharf, and gathered around their prisoner.

"This will never do," said Mad Twinker, shaking his head. "We can't take him through that crowd of people on the wharf."

"That's so," replied Buck Lamb, who had lost some of his prestige since his overthrow, and Mad Twinker had come to the front. "Four of us can handle him well enough now. Jeff Monroe, and you, Steve Douglas, go and get Bissell's four-oar boat."

"Bissell is n't there now," replied Jeff.

"Stop at his store as you go along, hire it for a couple of hours, and get the key," continued Mad Twinker. "Pull up to the creek at the head of the bay, and we will meet you there."

Though there appeared to be no one in the streets, for all the idlers had gone down to the wharf, there were people in the houses. Among the latter was Miss Susy Wellington, who had pulled an oar in the Marian. She had gone home to change her wet dress for a dry one, and saw from her chamber window the capture of Paul Bristol by the ruffians. She had heard something in the boat about Paul's battle with Major Billcord and his son, and she had some idea of the occasion of the assault upon the son of toil.

While Mad Twinker and his fellow-ruffians were hurrying their prisoner to the creek, she hastened down to the wharf. On her way she stopped at Bissell's store, where she had worked with Lily

Bristol. She learned that her employer had gone off in the sloop again, and that some of the Chesterfields had just hired the four-oar boat. She continued on her way, and soon saw the two students in her employer's boat, pulling rapidly towards the head of the bay. When she reached the end of the wharf she waved her handkerchief to the Marian, and beckoned with all her might with her hand.

Dick Short gave the order " Stern all ! " and the barge backed up within hail of the fair oarswoman. The coxswain asked her if she desired to pull an oar in the barge again.

" We are waiting for Paul Bristol, and he ought to be on board by this time," added Dick. " If he don't come, you can take his place."

" No, I thank you, Mr. Coxswain," replied Miss Susy, and she proceeded to detail what she had seen from her window.

" Paul a prisoner ! " exclaimed Dick, thoroughly aroused by the intelligence. " Stand by ! Give way lively ! "

He ran the Marian under the stern of the Gildrock, and reported the astounding intelligence to Commodore Chester.

" A prisoner ! " exclaimed Oscar, startled by the news. " Where is he now ? "

"Do you see that boat near the head of the bay? Two of the six fellows who captured Paul were sent for the boat, while the others took him in that direction."

"This looks like serious business. The Chesterfield barges both lie between us and that small boat," continued Oscar Chester.

The Gildrock was hauled around so that the commodore could confer with Thad Glovering. The details were repeated so that all the students in the three barges heard the whole story.

"It looks as though the rascals were going to bring Paul off in that boat, and put him on board of the Racer or the Dasher," which were the names of the two Chesterfield barges.

"Then, if we pull up there in a hurry, we may be able to get him away from them," added Thad Glovering.

"On the other hand, if the fellows in charge of the prisoner see our barges coming, they will put Paul ashore."

It was a difficult point to settle.

CHAPTER XIX.

A DEADLOCK AT THE HEAD OF THE BAY.

IT was finally decided by the commodore that all the Beech Hill barges should proceed to the head of the bay, and if the ruffians landed with their prisoner, or did not put him in the boat, they should follow them on shore and rescue Paul at all hazards. Oscar Chester gave the word, and the Gildrock dashed off, with the other two barges following her. The crews were not spared, and the boats appeared to leap over the water, which was tolerably smooth under the lee of the land.

The Chesterfield barges still lay near the shore, above the wharf, and in order to avoid them the commodore headed the Gildrock across the bay. In the absence of the spare hand and Dory, the Marian was one oarsman short, and Dick Short missed the stroke very much. The Silver Moon was not more than a quarter of a mile from the wharf, for Dory was showing off the sloop in various points of sailing. Very likely he desired to keep the Beech Hill barges in sight as long as the Chesterfields were near.

Dick Short waved his handkerchief in the direction of the sloop. Dory saw the signal, and headed the Silver Moon to the head of the bay. Before the commodore changed the course of the fleet, he ran across the stern of the Marian.

"The Chesterfields have made Paul a prisoner!" shouted Dick Short.

Dory heard the announcement, and then the sloop passed out of speaking distance. He was not a little astonished at the information, and fully realized the peril of the son of toil. Glancing at the Chesterfield barges, he saw Major Billcord seated in the stern-sheets of the Dasher, and it did not need a very vivid imagination to comprehend the programme of the enemy.

"I must ask you to excuse me to-day, and I will come over some other time when the wind blows, and put the Silver Moon through her paces," said Dory to the owner of the sloop.

"What's the trouble here?" asked Bissell.

The acting skipper explained the situation to him, including enough of the proceedings at Sandy Point the day before to enable him to understand it.

"Major Billcord is as savage as a wild hyena when he gets mad," added Bissell. "He is bound to have his own way against everybody else.   He tries to rule the town, though most of the people hate him."

Dory tacked and stood back to the Marian.   He hailed the coxswain and asked to be taken on board.   Dick told him to come alongside as quick as he could, and he would be ready for him.

" I wonder if I can't do something to help you," said Bissell.   " You fellows have done me a good turn to-day, and I should n't mind helping you out if I could."

" You will only offend Major Billcord, and I think we can manage the affair," replied Dory.

" No matter whom I offend ; Paul worked well in the sloop, and he seems to be the bottom dog in this business, and if I see a chance to do anything for him I shall do it, if the major bu'sts over it."

By this time the Marian was close under the bow of the Silver Moon.   Dick checked the head-way of the barge, and Dory put the helm down. Up went the starboard oars on the Marian, and the sloop was alongside of her the next minute. Dory leaped into the stern-sheets, and took his place at the stroke oar.   The sailboat fell astern, and the crew gave way again.

Dick gave Dory all the information he had in

regard to the capture of Paul. Both of them
were satisfied that the poor fellow would be
beaten half to death if he was not rescued from
the enemy. By the time the story had been told,
the Gildrock changed her course, and pointed her
bow for the creek at the head of the bay. On the
shore they could see several young men in the
uniform of the Chesterfield barges, and they could
be no other than the captors of Paul. The four-
oar boat was just making a landing.

When the commodore changed his course, the
barges of the two schools were at about the same
distance from the mouth of the creek. Oscar's
strategy had given Beech Hill this equal advantage. The movement of the barges created a
sudden sensation on board of the Racer and the
Dasher. Major Billcord began to demonstrate,
and a good deal of violent jawing came from the
crews. The magnate had seen the two ruffians
take the boat from the wharf, and pull to the head
of the bay. He could not fail to understand that
the prisoner was to be brought off in her.

Doubtless, he commended the prudence of the
ruffians in avoiding the crowd on shore. He had
kept faith with his hired villains for his own sake
as well as for theirs, and he was the only person

in the barges who expected Paul to be put on board, or who comprehended the movements of the six absentees from the boats. It was evident to him that the Beech Hillers had discovered what was going on, though it was a mystery to him how they had obtained their information, for the coming of Susy Wellington had not been observed.

"Run for the head of the bay, Jack Woodhorn!" exclaimed Major Billcord, when he saw the Beech Hill barges headed in that direction. "You must get there before those rascals from the other side do, or they will thrash the six boys of the institute who are there."

"We can't do anything, sir; we are four oars short in this boat, and two in the other," replied the coxswain of the Dasher.

"Don't waste a second, Jack!" protested the magnate. "Do the best you can. You have the inside track, and you ought to beat them with half a crew.'

Woodhorn gave the order to give way, and the eight rowers in his boat were soon pulling with all their might. The Racer followed her, and, having ten oars, she passed her. It looked like a race between the two schools, though it was a very unequal one. The Chesterfield students had improved in rowing a great deal since the last season, but discipline was still the wanting element in their organization, and though they had never measured speed with the Beech Hill boats, they were no match for them.

The boys from the other side did not seem to hurry themselves, but only pulled a steady and strong stroke. In five minutes it was clear enough that they were beating their opponents. The magnate urged the Chesterfields to greater exertion, and did more harm than good by his ill-timed interference.

Dory had shaken out the reef in the Silver Moon, and made an additional change in the ballast, so that she was now behaving remarkably well. Bissell had run over to the north side of the bay, and now had a slant which would carry him to the mouth of the creek. Oscar Chester had kept his gaze fixed on the party on the shore. He saw the six ruffians, and recognized Paul Bristol with his arms still bound behind him. The four-oar boat lay at the mouth of the creek, but the six ruffians had retreated to the high ground in the rear of the landing.

The approach of the three Beech Hill barges had completely upset the calculations of the ruffians. They stood looking down upon the lake, and appeared to be entirely non-plussed. The Gildrock was bearing towards the Westport side of the bay, and was coming between the shore and the Chesterfield barges. It was plainly folly to put the prisoner into the boat that had come for him. In the barges there were thirty-five Beech Hillers, and only twenty belonging to the institute. Besides, the boats from the other side had always been victorious over their own.

The Gildrock came to a stand, with the crew lying on their oars, and the other two barges followed her example at the order of the commodore. If the Chesterfields advanced, they would have to break their way through the Beech Hill line of boats. Jack Woodhorn ordered a halt before he came up with the formidable line in front of him. Colonel Buckmill, the principal, who was not present, had told the students of the institute never to come in collision with any of the boats from the other side, and the coxswains were disposed to obey their orders, especially as all the chances were against them.

"What are you stopping for, Jack Woodhorn?" demanded Major Billcord, when the oarsmen in the Dasher brought their blades to a level.

"We can go no farther without running into those barges," replied the coxswain.

"Run into them, then! Smash them if they don't get out of your way. Are you afraid of those chip-makers?" blustered the magnate.

"The principal ordered us never to come in collision with any of the boats from the other side, sir," replied Woodhorn respectfully.

"Are you going to leave your fellow-students on the shore to be mauled by those rascals?"

"I don't believe the tinkers will meddle with them as long as they stay on shore."

"But you want the rest of your oarsmen, and I want you all at Sandy Point. There is the biggest pile of fun for you over there that you ever had in your lives," continued Major Billcord, moderating his tone a little when he found his own wishes were in conflict with the orders of the principal.

Neither Jack Woodhorn nor Phil Fessenden, the coxswain of the Racer, was disposed to get into a row with the Beech Hillers. Both of them had been in the barges the summer before in all their tilts with the Gildrock and the Winooski, and they had learned wisdom from experience. It was in vain, therefore, that Major Billcord coaxed and threatened them. With a pair of black eyes out of the battle of the day before, Walk was hardly inclined to support his father, though he was quite as anxious as the magnate to get Paul into their possession.

For full a quarter of an hour the boats remained in the same relative position. The six ruffians on the shore had come to the conclusion that there was no getting out of the deadlock, and that the only way for them to earn their money was to march their prisoner to Sandy Point by land, a distance of two miles.

Bissell had run the Silver Moon into the mouth of the creek, and had been waiting for some movement on the part of the combatants in which he might do something to serve his friends. The ruffians were jawing among themselves as to what it was best to do, but he could not hear enough of their talk to understand their plans, if they had any. The skipper's patience was exhausted, and, taking his painter in his hands, he went on shore. Securing the rope, he walked up the bank.

"What are you trying to do?" he asked, addressing his remark to Mad Twinker.

"We want to put this fellow on board of the Dasher," replied the leader, as he had been since the overthrow of Buck Lamb.

"Well, why don't you do it?" asked Bissell briskly.

"Because the tinkers will interfere."

"I will take him in the Silver Moon if you like," added the skipper indifferently.

"Will you take the rest of us too?" asked Mad.

"Yes; I can carry a dozen well enough," replied Bissell.

About all the students were small customers at Bissell's store, and were well acquainted with him. They had no suspicion of any treachery on his part.

CHAPTER XX.

THE REBELLIOUS SKIPPER OF THE SLOOP.

WHAT are you going to do with Paul Bristol, Mad Twinker?" inquired Bissell, as they marched him down to the sloop.

"We are not going to do anything with him. Major Billcord wants to see him, and we promised to take him over to Sandy Point for him," answered the chief ruffian.

"Oh, that's all, is it?" added the skipper. "What is the major going to do with him?"

"We don't know; and it's none of our business."

"Of course it isn't," replied Bissell cheerfully, as he hauled up the bow of the boat so that the party could get on board. "How are you feeling now, Paul?"

"I think I am all right. I have got some hard cracks since I saw you at your house, but I guess I shall come out of it all right," replied the prisoner, looking with interest and anxiety into the face of the store-keeper.

Just then, while the ruffians were picking their way into the boat, Bissell gave the prisoner an almost imperceptible wink, which Paul saw and comprehended. It was full of hope to him, for he did not see how the skipper could deliver him over to the magnate after the good service he had rendered, in his humble way, on board of the Silver Moon. Besides, he was a Beech Hiller now, and the store-keeper knew it. He was under great obligations to them, and Paul did not believe he would betray one of their number.

The skipper had not lowered his mainsail when he made the landing, and the sloop was all ready to shove off. After two of the ruffians were in the standing-room, the prisoner was conducted on board between two others. At this point, Bissell went on board and took a stand near the tiller.

"It blows like Sam Hill to-day," said he, "and I want you to keep your places, and not move out of them. There are eight of us now, and sit four on a side. Here, Paul, you sit there," and he shoved the prisoner into the place next to his own, on the port side.

"But you won't have to go out into the rough water to get to Sandy Point," suggested Mad Twinker.

"We must go out some distance, for there is hardly any wind under the bluffs," replied the skipper. "You take a seat in that corner, Mad;" and he crowded him into the place opposite his own.

The other ruffians were arranged to suit him, and then he shoved the sloop off into deep water. The sail filled on the port tack, and the Silver Moon went off with the wind a little abaft of the beam. The shore was low at the head of the bay, and the sloop got her full share of the breeze. She struck into an eight-knot speed at once.

"It was lucky for us that you came up to the creek, Mr. Bissell," said Mad Twinker, as the boat shot ahead.

"Perhaps it was," replied the skipper; but there was not much enthusiasm in the remark.

"Those villains from the other side blocked us in so that we couldn't do anything, and we were thinking of walking Paul over by land," added the leader. "But some of us are about used up, and we did not like the idea of such a tramp through the woods."

As he spoke he glanced at the battered faces of some of his companions. They all looked as though they had been through the wars.

"The head boat of the tinkers is swinging around," said Alf Sumner, as the Gildrock turned her bow towards the shore.

"I wonder what they are going to do now," added Mad Twinker, with no little anxiety in his expression.

"The rest of the tinker boats are following her," continued Ham Jackson.

The Gildrock made a graceful sweep before the sloop came up with the position of the last barge in the line, and was abreast of the Silver Moon about as soon as she was under full headway. The Beech Hillers now laid themselves out, though they could hardly expect to keep up with the sail-boat in that wind.

"Can't you outsail those barges, Mr. Bissell?" asked Mad Twinker.

"Every time when we have as much breeze as we have now," replied the skipper. "And we shall have a good deal more before we have any less."

"There goes the Dasher," said Alf Sumner. "The Racer is after her."

"And both of them will be a long way after the tinkers," added Ham Jackson.

"We want you to put us ashore on the Sandy

Bay side of the point," said Mad Twinker. "Of course, we shall pay you the dollar an hour for the boat and boatman for all the time we have her."

"That's all right," answered the skipper, as he headed the Silver Moon farther out into the bay, and let off the main sheet to suit the change.

In a few minutes more the boat was in rough water, and she began to pitch and roll in a manner somewhat trying to the nerves of persons not used to it. The six ruffians, who were no boatmen, for they had very seldom been allowed in a sailboat, did not like it.

"What's the use of going out so far from the shore, Mr. Bissell?" demanded Mad Twinker.

"I don't think it is safe to sail near the bluffs, for the wind is flawy and snappish there," replied the skipper. "I don't know but I shall have to put a reef in the mainsail, for the gusts come heavier than I thought for."

As he spoke he hauled out a lot of rope from the locker under the tiller. He began to fuss over the lines to find a reef pendant. He took his knife from his pocket, and cut one of them off the right length. He laid the knife down by his side on the seat, and then returned all the ropes, except the one he had cut off, to the locker.

"I may not want to reef, but it is best to be ready," continued the skipper, shifting the tiller a couple of notches on the comb. "Do you think those barges will come up with us, Mad?"

"I should judge that they would not," replied the leader.

"The Beech Hillers are putting in some strong strokes," added Bissell.

"So are our boats," replied Mad.

"They are getting up quite a smart race. Will you fellows bet on your own boats?" asked the skipper, with a cheerful smile.

"Of course we won't while the Dasher is four hands short of her complement," said Jeff Monroe.

"But your boats are doing their prettiest, and I shouldn't wonder if they got the best of it in the end. Don't you see that the Dasher is gaining on the Marian?" continued Bissell, with a great deal of earnestness.

"I don't think she is gaining at all," put in Steve Douglas.

But the six ruffians were gazing with all their eyes at the five barges; and this was the one thing that Bissell most desired. While he held on to the tiller with his right hand he had picked up his knife with the other. Reaching around behind him, he got hold of the cord which bound Paul Bristol. Making sure that the blade was in the right place, in which he was assisted by the prisoner, he cut the rope.

"Is the Dasher gaining anything, Mad?" he asked when he had accomplished his purpose without attracting the attention of the ruffians.

"Not a hair; she is losing, and the tinkers are running away from our boats. They ought to when our fellows are short-handed."

"I guess you are right, Mad," added the skipper, as he looked about him, as if in search of something. "The wind comes stronger and stronger, and I think I shall want my long tiller. It is in the cuddy forward; will you hand it to me, Jeff Monroe?"

Jeff produced the spare tiller, and passed it astern to the skipper. It was about three feet long, and was made of the toughest oak. Bissell took it, and placed it at his side, between himself and the prisoner. Though Paul knew that his arms were free, he had not removed them from the position in which the cord had kept them, and no one but the skipper suspected that he was not still in bonds.

All that the owner had said about rough seas had been uttered to blind the six ruffians. It was rough, but not nearly so bad as it had been in the forenoon farther from the land. Bissell had become more interested that day in sailing a boat than he had ever been before. He had obtained a good many new ideas on the subject, and was really desirous of reducing them to practice. Without saying anything about his intention, he had gradually let off the sheet, and put up the helm until the Silver Moon was now a full mile from the shore, and was exposed to the entire force of the moderate gale.

The Beech Hill barges followed the sloop, but the Chesterfields were inclined to keep near the shore. The latter were short-handed, and this was doubtless their excuse. The sailboat was now at least half a mile from the Gildrock. Bissell was glowing with his new ideas, and he was disposed to profit by the instructions of the skipper of the Goldwing while they were fresh in his mind. Suddenly he hauled in the sheet, and threw the sloop up into the wind and then let her off on the starboard tack. Laying a course which would take him back to the mouth of the creek, he trimmed the sail and let her drive.

"What under the canopy are you doing, Mr. Bissell?" demanded Mad Twinker angrily.

"I am afraid the Dasher will not catch us if I run off any farther," replied Bissell.

"No matter whether she catches you or not. All you have to do is to land us at Sandy Point, on the bay side," added the leader of the ruffians.

"I guess we had better run back a piece," said Bissell, unmoved by the wrath of his passengers.

"We don't wish to go back," protested Jeff Monroe. "Do you want the tinkers to board us and take our prisoner out of the sloop?"

"I don't know that I care if they do."

"Don't you? Well, we do! We won't stand this sort of thing. We hire the boat, and she must go where we say," replied Jeff, rising from his seat, boiling over with wrath. "We won't stand it!"

"What are you going to do about it?" inquired Bissell in the mildest of tones.

"I order you to come about and take us to Sandy Point, as you agreed to do."

"I did n't agree to do anything of the kind. I told you I would take Paul on board, and then the rest of you, when you asked me to do so. That's the whole of it."

" This is treachery," yelled Mad Twinker.

" Well, if it is, I think we have carried this thing about far enough. I had n't any idea of helping you to hand Paul over to Major Billcord. I would sink my carcass to the bottom of the lake first," continued the skipper of the Silver Moon, warmly.

" All we have to do, fellows, is to take possession of the boat," shouted Jeff Monroe, as he made a rush aft.

" Back into your seat, or I will spill you into the lake ! " said Bissell sharply.

But Jeff was not to be intimidated by a threat, and, supported by Mad, he made a dive at the skipper. Suddenly the spare tiller appeared in air, in the hands of Paul, and then it came down upon the head of Jeff Monroe.

CHAPTER XXI.

AN OUTRAGE IN THE STANDING-ROOM OF THE SLOOP.

JEFF MONROE fell all in a heap at the door of the cuddy; but his fall did not deter his fellow-ruffians from advancing upon the skipper. Buck Lamb threw himself in front of Mad Twinker, as though he intended to redeem his escutcheon from the stain of defeat. Bissell was no more a fighting character than Paul, and he had not the least idea of "science." Buck was in proper position to make or receive an attack, and the skipper had risen from his sitting posture when the prisoner did.

Instead of hitting as a pugilist should, Bissell reached out his long arm, and took the bully by the collar of his uniform, jerked him over once, and then tossed him upon the prostrate form of Jeff Monroe. Mad Twinker had bravely followed up the attack until he was in reach of the skipper's arm, and he was tumbled over in a heap.

"His fall did not prevent his fellow-ruffians from advancing upon the skipper." — Page 216.

But Bissell could not do duty with his long arms
and steer at the same time; and the Silver Moon,
now having a strong weather helm, came up into
the wind, and, with her boom shaking in the midst
of the combatants, began to roll as though she
intended to pitch the ruffians overboard without
any help from her owner. Jackson, Sumner and
Douglas had retreated from the after part of the
standing-room, and the motion of the boat, made
more unsteady by the movements of the ruffians,
pitched them all into the seats.

Buck Lamb and Mad Twinker rose to their feet
as soon as they could, but Jeff Monroe was not yet
in a condition to move. The skipper put the helm
up, and the sloop filled away again. Paul had
advanced a pace, and taken a seat near the skipper, but with the spare tiller ready to deal a blow
as soon as a head came near enough to receive it.

The ruffians looked at the heavy tiller in the
hand of Paul, and then they looked at the long
arms of the skipper. While they were gazing
there was a halt all along the line, which afforded
an opportunity for reflection. Some of them cast
their eyes about them for something in the shape
of a weapon. The spare tiller seemed to be the
only stick that would answer the purpose of a club,

except the crutch used to support the boom when the sail was furled, and that was under the owner's seat.

Both Mad and Buck seemed to realize that they could do nothing without bringing that tiller down upon their heads, and its descent was almost sure to reduce them to the condition of Jeff Monroe, who was just beginning to show some signs of life. While they were thinking about it, the skipper came about, and headed the sloop towards the shore. When he had done so, he picked up the crutch, and placed it on the seat, by his side.

The Silver Moon was now headed in the direction of Sandy Point. The Beech Hill barges were some distance astern of her, and the Chesterfields at least a quarter of a mile farther to the westward.

"I don't think it is quite safe, Paul, to leave these fellows lying round loose in the standing-room," said Bissell, when he had the boat well in hand on the new tack. At the same time he drew out from the locker under him the ropes from which he had selected the reef pendant.

"What do you mean by that?" demanded Mad Twinker, before Paul had time to reply.

"I mean that I shall be under the disagreeable necessity of tying your arms behind you, just as you had Paul when you brought him on board," replied Bissell; but his tone and manner did not indicate that the duty was a very unpleasant one.

"But we won't stand it!" replied Mad angrily.

"Won't you?" continued Bissell, as he took the crutch in his hand.

"I protest against such an outrage!" exclaimed Ham Jackson.

"Oh, you protest, do you? And you really think it would be an outrage?" said Bissell, in a mocking tone.

"We are students in the Chesterfield Collegiate Institute, and we are the sons of gentlemen," returned Ham, with dignity. "Of course, it would be an outrage to put us in bonds, like common felons."

"Precisely so; then we will tie your hands behind you like uncommon felons. We won't quarrel about words, for you can beat me every time in that line. I suppose it was not an outrage for you to attack Paul Bristol, six to one, in the street, and tie his hands behind him," replied the skipper, with a very heavy sneer. "He don't happen to belong to the Institute, but he is a member of the Beech Hill School."

"We don't care what he is," interposed Mad Twinker. "We won't submit to be bound! We won't stand such an indignity!"

"You think you won't? I don't believe I shall have much trouble in tying the hands behind him of that fellow on the floor," continues Bissell. "It is for you to decide whether you will have your hands tied before or after you are knocked stiff. I will begin with you, Mr. Twinker."

"No, you won't!" cried Mad, as he sprang from his seat, and rushed madly upon the skipper.

Bissell was on his feet, and, reaching out his long arm, he took his assailant by the throat, in spite of the wild blows he aimed at him. This time he did not pitch him on the floor, but choked him till the ruffian began to feel weak, and to relax his struggles.

"You take the tiller, Paul, but don't let go the spare one. If one of them moves aft, knock him as you did the first fellow," said Bissell, as he picked up the reef pendant he had cut off.

Paul took the helm. He had steered a sailboat before, though he knew next to nothing about the general management of the craft, and kept her as she was. By this time Mad was decidedly faint, and the owner had no difficulty in tying his arms

behind him.   When he had done so, he picked up
the crutch again, and stepped forward.   His next
victim was Buck Lamb; but as about all the vim
had gone out of the bully, it was an easy job to
secure him.   The other four ruffians made no
resistance worth mentioning, for the crutch in the
hand of the skipper was an awful weapon; so was
the spare tiller which Paul wielded.   If the ruffians
could have got hold of anything in the way of
clubs, doubtless they would have held out longer.

As it was, they were completely vanquished.
Bissell had ranged them three on a side as he
bound them, and they had not been inclined to
move.   Possibly they thought they were in a bad
condition to save themselves if the Silver Moon
had taken it into her head to upset and spill them
into the angry lake.

"We are all right now, Paul.   Don't you say
so?" said the skipper, as he seated himself on the
weather side and took the helm.

"I should say that we were," replied the late prisoner heartily.   "You have saved me, Mr. Bissell,
from an awful pounding at the hands of Major
Billcord and his son; and I owe you a thousand
thanks.   I shall never forget what you have done
for me to-day."

"I think I got saved myself this forenoon," added the skipper; "and I guess I know how it feels. I think we should all have been drowned in the Silver Moon if it had n't been for the Beech Hillers; and I don't feel as though I could ever do half enough for them. We are not square yet, Paul, and you did your share."

"I did n't do much; it was Dory Dornwood who did it all," answered Paul modestly.

"It cut me like a sharp razor to give in to a mere boy on sailing, but Dory knows more about a boat with his eyes shut than I do with mine wide open. He did n't put on any airs, either," continued Bissell, with enthusiasm.

"By the big wooden spoon, there comes the Sylph!" exclaimed Paul, as the steam yacht came out from behind Barber's Point. "Captain Gildrock is on board of her, and you may be sure he will put things to rights in double quick time."

"All right; I am glad he is coming, for I don't know what to do with these fellows, now that we have them where they can't set the lake on fire." added the skipper, as he glanced at the Sylph. "We will keep on as we are, for I suppose she is going over to Beaver River."

"I think you have carried this thing about far

enough, Mr. Bissell," said Mad Twinker, after he had taken a glance at the steam yacht.

"Not quite, Mr. Twinker," replied the skipper, with a smile. "I shall carry it about as far as that steam yacht, and then I don't care a button what becomes of it."

"You mean by that to hand us over to the principal of the Beech Hill School?" inquired Ham Jackson.

"I reckon it amounts to that."

"If you settle the thing that way, you will have to answer to Colonel Buckmill," added Mad Twinker.

"I am ready to answer to him. Do you think the principal of the Chesterfield school will justify you in committing such an outrage as you have put on Paul Bristol?" demanded Bissell indignantly.

"Then why don't you hand us over to Colonel Buckmill?"

"So I would if he happened along here at about this time. Perhaps the other principal will pass you over to Colonel Buckmill," suggested the skipper, with a sort of chuckle, as though he had his doubts on the subject. "Six of you set upon Paul in the most cowardly manner, and —"

"Only two at a time," interposed Buck Lamb.

"How was it, Paul?" asked Bissell.

The intended victim of the ruffians told his story in full, not omitting to mention the punishment he had bestowed upon his assailants. Bissell expressed himself very emphatically in regard to the cowardly character of the assault, and was glad Paul had defended himself till he was overpowered by numbers. The faces of some of the ruffians bore the marks of his hard fists, and they were probably booked for mourning eyes by the next day.

By this time the Silver Moon was off Sandy Point. Paul had watched with interest the movements of the Sylph. When first seen she was headed for Scotch Bonnet, which was her direct course when homeward bound; but she soon shifted her helm, as though she was going up to Westport.

"She is going to make a landing at the town," said Paul, when he noticed the change of course.

"I am sorry for that, for I thought she was coming over this way," replied Bissell.

"Captain Gildrock must see all the barges, and I am sure he will run over here as soon as he makes them out," added Paul.

The words were hardly out of his mouth before the head of the steamer was turned in the direction of the Beech Hill barges, which were not half a mile astern of the Silver Moon. The skipper decided to come about, and stand back to the fleet. Before he could reach the boats, the Sylph had stopped her screw, and was hailing the Gildrock. Oscar Chester informed him that Paul Bristol had been taken a prisoner by six of the Chesterfields, and that the skipper of the sloop was conveying him to Sandy Point. This was all the commodore knew about the matter, but it was enough for the captain. He started the screw again, and in a few moments he had overhauled the Silver Moon.

From the pilot-house the principal could see the condition of things in the standing-room of the sloop. Again Captain Gildrock rang one bell, and then two. As the steamer lost her headway the Silver Moon rounded to under her lee side, where she had still water.

CHAPTER XXII.

AN INVITATION TO SANDY POINT.

I SHOULD like to hand these fellows over to you, Captain Gildrock," said Bissell, hailing the principal. " We have tied their hands behind them, and they won't set the lake on fire just yet."

"I should like to know something more about the case. Will you send Bristol on board to report to me?" replied Captain Gildrock. " Mr. Wolfenden, of Westport, is on board, and will join you in the boat."

"Wolfenden! Why, he is the deputy sheriff!" exclaimed Bissell. "I guess he is the very man we want."

"Mr. Bulfington is also on board," added the principal.

" Perhaps both of us had better go on board of the Silver Moon," said Mr. Wolfenden, for both he and Bulfington, who was a constable, were in the pilot-house.

"I should like to have one of you remain and hear the statement of Bristol, upon whom this outrage has been perpetrated," added Captain Gildrock. "As Bulfington is the constable, perhaps he had better remain."

The deputy sheriff was satisfied to let it be so, and he went on board of the sloop. The two officers had been down to Port Henry on official business, and, manifesting quite an interest in the Sylph, the principal had invited them to take passage to Westport in her. As the officer went on board of the sloop, Paul left her, and hastened to the pilot-house.

"Well, Bristol, it seems that you did not see the last of the trouble at Sandy Point yesterday," said the principal, bestowing a look of kindness and sympathy upon the new pupil. "Your face looks as though you had had a hard time of it to-day."

"The toughest time I have had yet, but not so tough as it would have been if Mr. Bissell had not stood by me, and got me out of the scrape. We did him a good turn this forenoon, and he did not forget it," replied Paul.

"I am glad to see you, Paul," added Mr. Bulfington, taking him by the hand. "It is about time the pranks of those students should come to an end; and I think the people of Westport have had about enough of them."

"What service did you render to Mr. Bissell?" asked Captain Gildrock.

"It is rather a long story, sir;" but Paul proceeded to relate it in full, and he soon found that both of his auditors were deeply interested in it.

"My daughter was one of the six girls," said Mr. Bulfington, when the spare hand had finished his narrative. "I am sure I owe the boys a debt of gratitude which I shall never expect to discharge."

Paul then gave the particulars of the assault upon him very minutely, and described the events which had followed his capture up to the present moment.

"It would have gone terribly hard with you, Paul, if Major Billcord had got you over to the point, for there is no more reason or mercy in him than there is in a fighting bulldog," added the constable. "He has got money enough to pay all damages, and he would not mind a thousand or two if he got his revenge in full."

"These boys must have been employed to capture Bristol," said the principal, whose face was flushed with indignation. "Mrs. Bristol this morning gave her son into my charge, and he shall not be abused while he is in my care."

"What do you intend to do about it, Captain Gildrock?" asked Mr. Bulfington.

"I shall prosecute the ruffians first; and if I can prove that Major Billcord employed them to assault my pupil, I will prosecute him," replied the captain.

The principal went to the side and had a brief talk with Bissell. It was arranged that the skipper should convey the boys to Westport, the constable going with him. Mr. Wolfenden returned to the steamer, and instructed the captain in the proper method of procedure in the court.

Bissell lost no time in getting under way. The ruffians were utterly crestfallen when they understood that they were to be prosecuted for the outrage upon Paul. Very likely Colonel Buckmill and the magnate would choose to regard it as a mere lark, a little wildness, on the part of the students, which ought to be passed over without any appeal to the courts.

Before this time the Beech Hill barges had come to the side of the Sylph. The Chesterfield barges had given the steamer a wide berth. They were close up to the shore, and were pulling in the direction of Sandy Point. Major Billcord could not help seeing his minions on board of the sail-boat, and to suspect that they had come to grief; but he could not prevail on the coxswains to go near any of the craft from the other side of the lake.

Paul remained on board of the Sylph, for his evidence was needed in Westport in getting out the warrant for the arrest of the ruffians. The principal gave no orders of any kind to the com-modore of the fleet, and he was left to do as he pleased. The students were consulted in regard to their wishes. Dory had gone on board of the Silver Moon as soon as the Marian came alongside of the steamer, and had obtained from Bissell all the details of the capture of Paul, and the sub-sequent events.

The Sylph stood over to the town, and the three boats locked together for a conference. The first thing was to hear Dory's account of Paul's adven-tures. Then they decided to wait until the Sylph started for home, for they were filled with curios-ity to know what might be done with the ruffians.

Just then they discovered that the Chesterfield

barges were lying on their oars off the southern arm of Sandy Point. They had some curiosity, and doubtless were more interested than their rivals. All the boats retained their positions for over an hour, when the Sylph was seen to leave the wharf. In a few minutes more she had crossed the bay, and stopped her screw near the Marian. The eight-oar barge was ordered to come alongside the steamer, and Paul was taken on board.

Of course they could not separate until the students had heard the news from Westport. In a few words the spare hand informed them that a warrant had been issued on the testimony of Paul and Bissell, and the six ruffians had been committed to the lockup. They were to be examined the next day, and the witnesses were duly summoned. In the presence of the magistrate Buck Lamb had broken down, and he declared that they had been employed by Major Billcord to capture Paul for the sum of twenty-five dollars. Two of the others indorsed this statement, and the principal had procured a warrant for his arrest, which was now in the hands of the constable. Captain Gildrock had procured the services of the best lawyer in the county of Essex to look after the business for him.

The news from Westport was very satisfactory, and the question seemed to be whether or not a man who had money enough to pay the bills could outrage a poor boy with impunity. Captain Gildrock's blood boiled, though it did not bubble, or otherwise manifest its condition.

The commodore gave the order for a start, and in a short time the barges came up with the Chesterfields, though they were a quarter of a mile farther out in the lake. The Sylph was hardly moving through the water, the principal doubtless holding her back to see that no trouble arose between the two schools. He took the precaution to run the steam yacht between the two parties, and soon found himself within hailing distance of the Dasher, with Major Billcord in the stern-sheets.

The two gentlemen were somewhat acquainted, and had occasionally met on the lake and at the bank in Burlington. As the Sylph went lazily along, the captain discovered a white handkerchief hoisted on a stick, and saw that the Dasher was pulling towards the steamer. He rang his bell to stop her, and she awaited the pleasure of the barge.

"Captain Gildrock, you and I have always been good neighbors, though we don't live on the same

side of the lake," said Major Billcord, standing up
in the stern-sheets of the barge. "I feel it my
duty to give you a friendly warning. I learn that
a young scoundrel by the name of Paul Bristol,
whose family I have charitably harbored on my
land without the payment of rent, came over to
Westport to-day with the students of your school.
He is a young villain, and I warn you not to trust
him."

"I had come to the conclusion that he was a very
good boy," replied the captain.

"You are utterly mistaken, sir!" protested the
major. "He assaulted both my son and myself,
for which I intend to punish him in the severest
manner. His mother and sister live on the point
here, in a cottage owned by the woman; and the
boy lived here before he went to Genverres, if he
has gone over there. I warned the woman to
move her house at once. She has not done it, and
I shall tumble the building into the lake. It will
make some sport for our boys, and I thought yours
might like to see the fun, and learn a good lesson
in the administration of human justice. I should
be happy to have your steamer and your barges
take position near the point, where you can see the
proceedings."

Captain Gildrock made no reply, and the Dasher pulled away without waiting for any. The two Chesterfield barges ran their bows into the sand in front of where the cottage had stood, and the Sylph, after whistling for the Beech Hill barges to approach, followed the Dasher. The barges from the other side pulled to the entrance of the bay, and lay upon their oars.

"Now is the time for the fun to begin, and we are invited to see it," said Dick Short to Paul, who sat by his side.

"I think there will be some fun, though it will not be what Major Billcord and his crowd came to see," added Dory.

On board of the Dasher, Jack Woodhorn had risen from his seat, after he had given the order for the oarsmen to boat their oars. All the students were busy attending to their blades. Woodhorn was evidently looking for the cottage; but he did not see it. Then the magnate stood up; then all the students in the two boats stood up, and then both barges were nearly upset by this folly, and the coxswains ordered their crews to be seated.

"Did I understand you to say there was a cottage here to be tumbled into the lake to illustrate the administration of human justice, Major Billcord?" called Captain Gildrock, who had placed the bow of the Sylph within a few feet of the stern of the Dasher.

"I don't understand this," replied the magnate. "The cottage was here yesterday, and it was quite impossible for the woman to move it. Send the young gentlemen ashore to see if they can find it."

The young gentlemen could not find it. The site where it had stood was smoothed over as nicely as though the building had never been there. The major said it was a great mystery.

"No human justice to-day, then?" queried the principal of the Beech Hill school. "Perhaps divine justice had got ahead of human justice in this instance, as it sometimes does."

"Do you know anything about it, sir?" demanded the major angrily.

"The cottage was removed to Genverres by the students of the Beech Hill Industrial School last night," replied the captain.

CHAPTER XXIII.

THE PROCEEDINGS AT SANDY POINT.

IT is to be regretted that Major Billcord set a very bad example to the Chesterfield students by using profane language when the Beech Hill principal " let the cat out of the bag." At first he was not inclined to believe the statement, for it seemed incredible to him that any company of boys should have been able to put the cottage on a boat and convey it across the lake.

Captain Gildrock was obliged to explain how the work had been done, before he would accept the solution of the mystery. The principal offered to take him over to the school grounds and show him the cottage if he had any doubts on the subject.

"It was a shabby trick on your part, after all," said Major Billcord, when he had got the explanation through his head. " I don't interfere with your affairs on the other side of the lake, and I don't know of any reason why you should meddle with mine."

236

"As I understand the matter, major, you warned Mrs. Bristol to remove her cottage within twenty-four hours, or you would tumble it into the lake," replied the captain. "I simply allowed the students to assist the good woman in doing what you required her to do."

"You knew very well that this was a case of discipline on this side of the lake," replied the major, waxing exceedingly wrathy. "My son was attacked in the most brutal manner by that woman's cub, and she upheld the young villain, and would not allow him to be punished as he deserved. Of course, I knew the woman could not remove the cottage, and it would have been here now if you had not meddled with my business, like a very bad neighbor."

"I don't care to discuss the assault, as you call it, with you, but I think the boy and his mother were fully justified in their action," replied the principal, in firm but gentle tones.

"You do, do you?" demanded the major. "The young rascal abused my son. Look at his face," and he pointed at the interesting features of Walk, who sat in the boat listening to what his father said. "When I interfered to save my son from serious injury, he flew at me like a wildcat, and look at my eyes."

"Why did n't you prosecute the boy in the court, and have him properly punished?" suggested the captain, looking rather sarcastically at the magnate.

"I don't do business in that way," returned the major, boiling over with anger. "I shall punish the young rascal myself! I shall do it here at Sandy Point, where the outrage was committed. I have taken steps to have him brought here for that purpose."

"Then you expect Paul to be brought here, do you?" asked Captain Gildrock, astonished at the remarks of the magnate, for he had not a doubt that he had seen what had transpired while the sailboat was alongside the Sylph.

"I do expect him here very soon; and I shall tie him up to a tree and give him such a thrashing that he won't get off his bed for one month after it!" exclaimed Major Billcord, flourishing a rawhide in the air as he spoke.

"This strikes me as rather brutal," added the principal.

"Brutal? Look in Walker's face! Look in my face! Were the blows that made these marks brutal, or were they not? I shall have satisfaction for them!"

Captain Gildrock was trying to explain to himself how it happened that the magnate knew nothing of what had taken place alongside the steam yacht. As he thought the matter over he understood it better. The Sylph had been between the Silver Moon and the Chesterfield barges, so that they could not see what took place on the lee side of her.

While Paul was telling his story in the pilothouse, the Chesterfields, finding the Beech Hill boats gathering around the steamer, had pulled close up to the shore, and continued on their way to Sandy Point. Neither Major Billcord nor the students with him had been troubled with a doubt in regard to the fidelity of Bissell to their interests. Even if he was faithless, he had the six ruffians with him, and they would be more than a match for a single man.

However it had happened, it appeared that the magnate and the crews with him knew nothing at all about the capture of the ruffians. They had waited off the point till the Beech Hill fleet came up, and they concluded that the sloop was standing off towards Westport to avoid a meeting with the " tinkers."

The magnate wished the visitors from the other

side to see the destruction of the cottage, and he had invited them to be spectators of the expected frolic. He had decided to attend of the cottage first, so that the Beech Hillers might see the fun, and to administer the punishment to Paul Bristol after they had gone, for he did not care to have them witness that spectacle.

"I am afraid you are laboring under a mistake, Major Billcord," continued Captain Gildrock. "Did I understand you to say that you expected Paul here to be punished for his brutality?"

"That is precisely what I do expect; and he will be here in the course of half an hour. But I need not detain you any longer, sir," replied the magnate, with very ill grace. "There is no mistake about it, you may depend upon it."

"If you will pardon me, there is some mistake, for Paul Bristol is here now," added the captain. "Marian, ahoy! Back down this way," he shouted to the barges, which were lying at the entrance to Sandy Bay.

The crews were lying on their oars, and Dick Short promptly gave the order, "Stern all," and the Marian soon put her stern very near the bow of the steamer.

"Stand up, Bristol, and show yourself," said the captain.

Paul stood up in the stern-sheets of the barge, and Major Billcord looked at him as though he had been a spectre from some neighboring grave-yard. Then he had a moment's animated conversation with the coxswain at his side. It was another mystery, and possibly the magnate thought he was in the middle of the last chapter of a novel. How had it been possible for him to get out of the clutches of the six ruffians?

But the mystery suddenly paled, and the major threw himself into a towering passion. The object of his intended vengeance was before him. Jack Woodhorn, at the request of the magnate, summoned his crews from the shore. It looked as though the enraged major intended to attempt the capture of his victim under the very eyes of the Beech Hillers.

"That is Paul Bristol, as you may see for yourself, Major Billcord," said Captain Gildrock, when he observed the preparations on the part of the Chesterfields to do something. "He is now a student in the Beech Hill Industrial School; and to him, for the time being, I stand in the relation *in loco parentis*; I shall protect him to the fullest extent."

"Captain Gildrock, this is unfriendly to me, and —"

"But friendly to the boy and his mother, who need a friend more than you do," interposed the principal. "The six ruffians you hired to capture that poor boy have been arrested and committed to the lockup. Some of them admitted that they were employed by you to do this piece of villany, and there is a warrant out for your arrest. Doubtless, the facts will all come out at the examination to-morrow forenoon; and if it appears that I have done you any wrong, I shall be prepared to make you abundant reparation."

"A warrant for my arrest!" exclaimed Major Billcord, sinking down into his seat.

"I have employed the Hon. Richard Lawbrook to look after the case in my absence, and I hope justice will be done," added the captain as he rang two bells in the pilot-house.

At the same time the principal made a motion with his hand, in the direction of home, to the boats ahead of the steamer. The barges backed into position, and the commodore shouted the order to give way. In a minute more they were in line, pulling down the lake, but keeping near the shore.

Major Billcord did not utter a word to anyone. He was in deep thought. Very likely his impul-

sive nature had led him to organize the plan for
the capture of Paul without any consideration of
the possible consequences. He was by far the
richest man in that region, and owned no end of
shares in all the industrial and commercial enter-
prises of that part of the State. He was a man of
large influence, and was not over-scrupulous in
regard to the use of it. With such power, he was
in the habit of having his own way, though there
were a few people in the neighborhood who contrived to maintain their own independence, even at
the risk of quarrelling with the magnate.

Among the latter was the Hon. Richard Lawbrook, a prominent lawyer in the county, who had
been made a senator, though without pledges, in
part by the influence of the major. But when the
legislator was requested and pressed to promote
by his eloquence a more than questionable enterprise, his conscience revolted, and he refused his
aid. This had produced a bitter quarrel between
himself and the magnate, though all the honest
people believed that the senator was an upright
and just man.

Doubtless, the mention of the senator's name had
produced a decided effect upon the mind of the
magnate. Mr. Lawbrook was a man of influence,

who believed that the laws should be impartially
executed upon the rich as well as the poor. The
prospect ahead was not pleasant.

There was no "pile of fun" to be had at the
point that day, and Commodore Woodhorn backed
away from the beach as soon as his crews were in
a condition to do so. Without asking the major
any questions, he conveyed his distinguished passenger over to the town, and landed him at the
steps. The procession was formed to escort him
to his elegant mansion when Mr. Bulfington
appeared, and respectfully announced that he had
a warrant for the arrest of the object of the parade.
The major was impatient when the officer presented
himself, and told him to call at his house if he had
any business with him.

The constable politely intimated that it was a
criminal proceeding, and that he was under the
necessity of taking his prisoner wherever he could
find him. He treated the culprit just as though
he had been a poor man, which was a new experience to the magnate. He was taken to the
lockup, and confined in a cell. With the major
behind the bars, Mr. Bulfington was complaisant
enough to do anything he required. A couple of
wealthy friends were sent for, and the major and

the six ruffians were bailed out in the course of an hour.

There had been some earnest talk between Captain Gildrock and Mr. Lawbrook, for the former wished to know whether justice represented a substantial idea in the State of New York. The senator was confident that all men were equal before the law; and as he had more influence with the constable than any other person had, Mr. Bullington was unwilling to assume any special responsibility in regard to his powerful prisoner.

The Beech Hill fleet went home, and not only the students, but the families at the mansion and at Hornet Point had enough to talk about for the rest of the day. The next morning, when all the students except Paul were at their studies in the school-room, the Sylph, with Mrs. Bristol and Lily on board with Paul, sailed for Westport.

CHAPTER XXIV.

THE STUDENTS DECIDE "WHAT'S IN A NAME."

MAJOR BILLCORD had the services of a distinguished lawyer from Plattsburgh and of another from Elizabethtown to defend himself and the six ruffians. They almost made a "celebrated case" of it. They got the trial into the county court, and the six ruffians were sentenced to ten days confinement in the county prison, and the major himself to thirty days.

The assault and the conspiracy were too outrageous, in the opinion of the justice, to be punished with a simple fine. The workingmen and the farmers had got hold of the case and talked a great deal about it. Against the advice of the eminent legal gentlemen he employed, Paul Bristol was prosecuted for the first assault upon Walker Billcord, and this brought in the testimony of Lily and her mother, which the lawyers wanted to keep out. This evidence completed the history of the case by filling in the beginning of the trouble.

246

Paul was fully justified and acquitted, and the people praised him for his defence of his sister. Fathers and mothers were interested, for they looked to the law for the protection of their children. Every effort was made to save the magnate from the disgrace of being confined in a common prison, but public sentiment would have been indignant, and he had to serve out his time.

The farmers and mechanics were inclined to go as far the wrong way in one direction as the major and his friends were in the other. The magnate was a "soulless capitalist," a "bloated bondholder," and he suffered, especially among the mill hands, for this senseless reason. But the judge was even-handed between the two parties, and the major learned a lesson which was worth half his fortune to him — that before the law he was no more than the equal of the poor man.

Four of the six ruffians were taken from the institute by their parents, who had sent them there to be fitted for college, and they were sure that their sons had been led away by the influence of Major Billcord, and by the lack of proper discipline in the school. Three others were removed for the same reason. The loss of these pupils was a severe blow to Colonel Buckmill, who had condemned the conduct of the major from the first.

Though other students were soon obtained to take their places, and even to increase the number of pupils in the school, the colonel realized that he was not managing the institution on the right principle. The magnate had lost much of his influence in the vicinity, and the principal found that he could afford to be independent of him, for it would pay better.

On the day of the examination, Captain Gildrock called the students of Beech Hill together in the school-room in the afternoon, after the studies had been finished. The model to which he had alluded at the time of his lecture on shipbuilding was still on the table where it had been placed on that occasion. The students had examined it with a great deal of interest. They had read all they could find in the books in the library on the subject, and studied the three plans on the wall.

They were very impatient, as young men always are, to begin the actual work of building the boat. It was even more to their taste than erecting a house, though many of them were now competent to frame a building from the plans.

The subject of a name for the craft had engaged

their attention, and they had given a great deal of thought to it. They were all ambitious to name the schooner, and a great variety of names was likely to be presented. They had all been handed in ; and when the principal announced that the first business was to select one from them, Mr. Bentnick handed him the envelope which contained them.

"Lily," said the captain, laughing, as he took the first paper from the enclosure.

All the boys looked very good-natured, though something like embarrassment appeared upon the faces of not a few of them.

"Lily," continued the captain, reading the second suggestion.

The good-natured looks expanded into smiles.

"Lily," the principal said again, as he drew out the third paper. The smile became a little more intense.

"Lily," repeated the principal once more, and then he emptied all the slips of paper from the envelope, and began to sort them over.

Principal, instructors and students were all laughing merrily by this time. It was evident that the boys were very impressible fellows, and had been captivated by the beauty of Miss Bristol.

Possibly some of them were disturbed because they found that others had made the same selection as their own.

"They are not all alike," said Captain Gildrock, when he had finished sorting the papers. "Only about two thirds of them are 'Lily.' It is certainly a very pretty name, and there is no flower more pure and beautiful than the lily. But the name is rather general and indefinite. We have the tiger lily, the lily of the valley, the pond lily, and other kinds. What do you say to calling the schooner the Pond Lily?"

"No, sir!" shouted a majority of the students, with one voice.

"The Tiger Lily, then?"

"No, sir," was the emphatic reply.

"Then Lily of the Valley?"

"No, sir!" again voted the majority.

"Perhaps I shall have to ask Miss Millweed for the names of other kinds of lilies," added the principal, with a very pleasant smile.

"None of them!" exclaimed the crowd, encouraged by the cheerful expression of the captain.

"None of them?"

"Lily Bristol!" called Luke Bennington. "I put in another name, but that is what the fellows mean."

"Yes, sir!" cried the majority.

Captain Gildrock improved this opportunity to say something about the influence of female society, and especially of young ladies. If Beech Hill were not a school of mechanic arts, he should be in favor of having as many young ladies as young gentlemen on its roll of pupils. He was in favor of co-education, whereat Mr. Bentnick shook his head, and seemed to be uneasy in his seat, though Mr. Darlingby showed a disposition to clap his hands. The captain was an old-fashioned man, he said, but he hoped he had modern and progressive ideas. He was not in favor of "pretty girls."

At this point about half a dozen of the students gave something like a suppressed groan. The principal paused, the dissentients wished they had said nothing; but he did not add a word. He seemed to feel that they had as much right to express themselves in this manner as to applaud, or express themselves in other ways.

"I don't believe in pretty girls as such," he continued, "because they monopolize the sole attention of young men, to the exclusion of others even more worthy who are personally less attractive. But I hardly expect young gentlemen to adopt

my views on this subject before they have lived
to be as old as I am. When you have an opportunity, boys, bestow some attention upon the
' wallflowers.' "

He had not intended to say so much on this part
of the subject, and he resumed the general topic.
Ladies should be treated with the utmost respect,
whether attractive or not, and even if they did not
conduct themselves like ladies. He did not pro-
hibit the students from associating with the young
ladies of Genverres and the neighboring towns,
under proper circumstances, and thought female
society was beneficial to them. But profound respect must be the basis of such relations. There
should be nothing like undue familiarity with
them, and a young lady, even if not more than
fifteen, should not be addressed by her Christian
name except by her relatives. It must always be
"Miss Bristol," and not " Lily," or even "Miss
Lily."

" Not Miss Lily ? " said a puzzled student.

" That is a grade of familiarity between the sur-
name and the given name, proper enough for persons who are intimate enough to use it, but not
applicable in the present instance. Now, to return
to the name of the schooner, from which the papers

withdrew my attention. I am entirely willing that you should give her a name."

After what the principal had said about pretty girls, most of the students concluded that any attempt to give the name of one of that sister-hood to the craft would be vetoed. They were not quite satisfied to have their wishes disregarded. His last words, however, gave them a little encouragement. The principal picked up the slips of paper and counted them, or a portion of them.

"Twenty-three have given in the name of 'Lily,'" said he, taking up those which indicated some other name. "'Champlain,' 'Lake Bird,' 'Lake Gem,' 'Saranac,' and the names of most of the lakes and rivers in Vermont. Among them are 'Addison' (our county), 'Genverres,' either of which would be a very good, and a very appropriate name. I don't like the fancy names, such as 'Gem of the Lake,' as well as the more substantial ones. Now you may vote on the question, and the name among those I have read which has a majority shall be the one selected."

This announcement brought out some applause. Then the captain said it would be in order for any student who wished to recommend any particular name, to say what he pleased on the subject.

This permission brought Luke Bennington to his feet.

"I don't believe there is a craft on the lake now called the 'Champlain.' There has been a large steamer, but she is no longer in existence," said the speaker, with considerable earnestness. "In my opinion — and I have seen the lakes of Scotland and Switzerland — Lake Champlain is the finest lake in the world."

"Oh! Oh! Oh!" murmured several listeners.

"Of course the size of the lake comes into the comparison, or Lake George might be nothing by the side of a little pond between a couple of hills. We have the Adirondacks on one side, and the Green Mountains on the other; and, taking everything into consideration, I vote for Lake Champlain as the finest. For this reason, and because it is the element on which the schooner is to figure, I hope its name will be given to her."

Another student spoke in favor of "Battenkill," but he proved to be the only one who had ever heard the name. He explained that it was a river which had its source in Vermont, though it soon wandered into New York. A third student spoke for "Addison," and a fourth for "Genverres." The advocates of "Lily" seemed to be conscious of their strength, and all of them were too bashful to make a speech in its favor.

Mr. Darlingby passed around the box, and all voted. The ballots were quickly counted by the instructors, who appeared to be quite as much interested as the boys, and the principal read the result. It was evident then that a few did not care a straw for the name, and voted for the fun of it.

"Tiger Lily, 1; Pond Lily, 1; Lake Gem, 1; Genverres, 1; Addison, 1; Battenkill, 1; Champlain, 2; Lily, 28," the principal read from the paper made out by Mr. Bentnick.

The vote was received with tremendous applause, and the students continued to clap their hands until the captain raised his finger as a signal for them to cease.

"The schooner will be called the 'Lily,'" added the principal.

CHAPTER XXV.

BUILDING THE BOAT.

WHILE the boys were getting over the excitement attending the vote on the name of the unbuilt schooner, Captain Gildrock moved the table on which the model of the craft stood to the front of the platform, where all could see it. It was a very graceful form, and the nautical boys were sure that the schooner would be a fast sailer.

"I told you before," said the principal, " that this model was constructed on a new plan ; but I have since learned that the idea is not as original as I supposed it was, and that boats have been built by this method. I supposed I should be the first to introduce it, but I find I am not, though the model-maker had never made one before.

"The plans are drawn and the model made on the scale of one inch to a foot. Whatever measures an inch in the plan or model measures a foot in the boat. Every stick in the craft will there-

256

fore be twelve times the size it is in this model.
The Lily will have a beam of thirteen feet, which
is a little more than one-third of the length.

"This model is made of soft pine and mahogany.
The ribs are to be twenty-one inches from the centre of one to the centre of the next one. What
looks to you like the space between the timbers,
or ribs, is pine. The mahogany strips, which are
comparatively narrow, represent the timbers. The
pieces are tranverse sections of the hull, made
separately and put together.

"This is the stem," continued the principal, de-
taching the piece from the model. "A model of
it is to be made of soft wood, enlarged to twelve
times the size of this piece. Removing a piece of
the pine, we come to a mahogany section. As the
bow is round, these sections radiate from a shorter
line on the keel than the horizontal line above it
on the deck would be."

The captain held up one of these diagonal sections, the top or deck part of which was in the
shape of a triangle with a very acute angle. The
parts were in pairs, one belonging on each side of
the bows.

"The outer edge of this half section gives the
shape of the bow where this piece belongs. If

you lay it on a smooth board, you can mark from it the exact curve of the timber which is to go where this piece is taken out. It must be enlarged to twelve times the size of the section. Of course the outer surface of the section only is of any use to you; but having obtained the exact shape of the outside, the rib may be made of any width and thickness we wish."

The principal removed several of the diagonal half sections, all of which were to be treated like the first one. At the widest part of the model he came to pieces which were of uniform thickness. These were also in pairs, separated on the vertical plane above the keel. The lines of the outer surface in these half sections were to be transferred and enlarged as before, and the mould or pattern was to be made from them.

In this manner the model was pulled to pieces, and from the edges of the transverse sections the shape of the craft was to be obtained. The stern-post, stern-knees and transoms were to be shaped in the same way. The captain stated that the model represented the craft after she was planked, and it would be necessary, in drafting the moulds, to allow for its thickness.

Mr. Jepson then took the platform and explained

in what manner the outlines of the half sections were to be enlarged. To give all the details would take a whole volume, and doubtless it would be very dry reading to most young people. Only an idea of how the work was to be done can be given, and those who wish to build such a craft as the Lily, or even a boat or a canoe of a third of her length, need full drawings and very minute explanations.*

These lectures used up the rest of the afternoon. The next day was devoted to making the moulds. At the end of ten days the frame was ready to set up. The keel was laid down at an angle of three-quarters of an inch to the foot, so that it would readily slide off when the time came to launch it. The boys had been studying on the subject, and the principal had supplied the library with all the available works. They labored very hard because they were very deeply interested.

Setting up the frame was an exceedingly exciting labor with them; but when it was up they found the shape corresponded with the model. Some

---

* The editor of "Forest and Stream" has prepared a work for amateurs, called "Canoe and Boat Building," which may be obtained of the publishers of that valuable paper, or at the bookstores.

had done their work better than others, and here
and there it was necessary to do considerable fit-
ting. In another week the frame was ready for
the planking. Captain Gildrock gave frequent
lectures on the proper methods of doing the work,
and explained the construction of larger vessels
than the Lily.

A part of the Champlain mechanics, as they still
called themselves, worked in the shop, and a part in
the building-shed. There were a great many bolts
of iron and copper, and a great many metal plates,
braces and straps to be prepared, which gave
abundant employment to the machinists, who had
been instructed by the head of this department in
forge-work, as well as filing and turning.

The carpenters had plenty to do in the shop,
with abundant opportunity to learn many things
which are not required in the ordinary experience
of such mechanics. There was also room enough
for the exercise of their inventive powers.

With so many enthusiastic workmen, who found
abundant variety in their operations as the schooner
advanced towards completion, the planking was
soon finished. Then a dozen of the carpenters
went to work upon it with the smoothing-planes,
and the outer surface was made as smooth as a
floor.

The hull rested in the cradle which had been erected for it, and the tops of the timbers were secured in their places by cross-stays. So far, little or nothing had been said about the interior of the hull, for the reason that the frame and planking had absorbed all the attention of the workmen. The next step was to put in the deck-beams, and secure the shelves on which they were to rest.

"Now, boys, we must decide upon the plan for the inside of the schooner," said the principal, when he had called the students together in the building-shed. "What accommodations shall we provide on board?"

"A cabin and a standing-room," replied Life Windham.

"Like the Goldwing," suggested Matt Randolph.

"The Goldwing contains a cabin, cook-room and standing-room. We can have all these on a larger scale in the Lily; and there will be space in the run for a store-room and ice-house, with a door into it from the cabin."

"Behind the steps at the companion-way," added Matt.

"There is no other way to reach it unless you put a scuttle in the floor of the standing-room, which is liable to leak," replied the principal. "The steps can be hung on hinges and turn up, but I think it is better to slip them back out of the way. How long will you have the cabin?"

"How much clear space have we inboard?" asked Luke Bennington.

"About thirty feet, after allowing for the bend of the bows and the rake of the stern," answered the principal.

"Cabin fourteen feet, I should say," continued Luke. "That will leave eight feet besides the overhang for the standing-room, and the same for the cook-room."

"That was my calculation," replied Captain Gildrock. "Then we want a trunk fourteen feet long, which may be about nine feet wide on the main deck. This will make a very roomy apartment for a lake craft. On each side of it we must build up transoms, or divans, for seats or berths. As we have no centreboard in the middle of it, there will be nothing in our way."

"Why did n't we build a centreboard boat?" asked Matt Randolph.

"I think a keel boat is safer and stiffer. In Lake Champlain we have plenty of water, though we shall draw about five feet aft. There are shoal places, but there is n't the least need of running over them."

" In a centreboard boat, if you get aground, there is a chance for you to work off when you cannot in a keel," said Matt.

" That is very true ; but I think the stiffness of the keel craft more than compensates for the advantage of light draft in these waters."

The principal then made a drawing to illustrate the method of putting in the beams and the knees that were to support them.   As only a portion of the beams could extend entirely across the boat, on account of the elevation of the trunk, it required careful work and planning to secure the necessary strength.   But this problem had been solved by the instructors, and the descriptions of the timbers were obtained.

In a few days more the frame of the trunk and deck was in place.   While a part of the workmen were planking the deck, the rest were putting down the floor of the cabin, and building the transoms.   Before the 1st of July the work on the hull was completed.   The boys had contrived a great many lockers in the cook-room and cabin, for the storage of dishes and cooking utensils, and for everything needed on board.

Inside and outside, all hands went over the work with sandpaper. A gang of calkers had already filled the seams with oakum. Tar, pitch, and putty had been used where they were needed, and no one considered it possible for the craft to leak a drop.

A thin coat of lead color was then put on the outside, and one of white inside. The boys had some skill in painting, for they had been called upon to do all kinds of work, from laying brick up to tinkering a watch. Several coats were given to the whole, but the last two on the outside were of black.

Captain Gildrock had decided to have an iron false keel added, partly to protect the wood and partly to serve as ballast. A pattern of the casting had been made and sent up to Port Henry to be cast. It was in one piece, and weighed over a ton. Of course, it had to be bolted on before the Lily was launched. It was too heavy and cumbrous to be transported on the Sylph; but Mr. Miker had to deliver a cargo of stone at Port Henry, and it could be brought down on the gundalow. It was nothing but fun for the students to tow the unwieldy craft about the lake, and the next Saturday holiday was to be used for this purpose.

At eight o'clock in the morning the Sylph, fully manned by the students this time, started out of the creek with her tow. Paul Bristol had been assigned to a place as a deck hand on board, and he had made several trips in the steamer. On the present occasion he had asked to be excused, in order to attend to some work for his mother.

For two weeks before, Lily had been at work for Mr. Bissell, taking the place of Susy Wellington, who had gone to visit her friends in Albany. She had written to her mother that she should like to spend Sunday at home. It was difficult and expensive to get from Westport to Genverres then, for the steamers· did not go up Beaver River.

Paul thought he could manage it. As it was a still day on the lake, he was going after her in the flatboat, in the afternoon. It was a ten-mile pull, but he was good for that. He had hauled up the boat after dinner to put it in order, when he discovered a queer-looking craft coming down the creek.

CHAPTER XXVI.

THAT CUNNING TOM TOPOVER.

THE strange craft appeared to be modelled after the catamaran, though the builder thereof had never seen one. It consisted of two logs a foot in diameter and ten feet long, which were placed three feet apart. In the middle of the supports two boards were nailed on the sticks, so that the thing looked more like the letter H than it did like a catamaran.

On the cross-boards was an old window-glass box; and on the box was seated Tom Topover. He was the naval architect of the structure which bore him, and the craft was not at all creditable to his ingenuity. If he had nailed the cross-boards across the ends of the logs, there would have been more stability to the affair, though in that case it would have been stigmatized as a raft.

In his hand Tom held a rude paddle with which he was trying to control the movements of the aquatic chariot. It had no propelling power, and

floated with the current down the creek. A bend
of the stream threw the force of the water against
Hornet Point, and Tom was trying to keep it
from going against the rocks. He did not succeed
at all, and one of the logs, striking the bank, twisted
the structure entirely out of shape.

The two logs came together, the nails which
held the cross-pieces twisting off with the slight
shock. Paul saw that Tom was in peril, and he
rushed into the house to get his oars, which he
kept in his chamber, in the attic, for he was afraid
that some of the hard-looking boys of the Topover
herd might steal his boat. He had heard of Tom
before, but he had never seen him, and he did not
know that the fellow on the queer craft was he.

Tom Topover dropped from the box down upon
the log, allowing his seat to fall into the water.
With the paddle in his hand he threw around the
end of the affair, till it was within reach of Paul's
flatboat. Reaching out, he hauled it in, and
jumped into it. The logs floated off with the
current of the little lake.

"You are there, are you?" said Paul, as he returned with the oars.

"Yes I am. How are you, Paul Bristol?" replied Tom, with a grin from ear to ear.

"Well, I thank you; how are you?" added Paul. "I thought you were booked for a bath, and I went in for my oars."

"That's right, and you are a good fellow. They say you are a Bristol brick."

"Perhaps I am, but I don't happen to know you, and can't say what sort of a brick you may be," replied Paul.

"I'm a perfect brick. Gi' me them oars and I'll pick up them logs," continued Tom, extending his hand to receive them.

"I'll help you do it," replied Paul, taking the painter and hauling in the boat.

"What's your name?"

"Jack Sheppard," replied Tom, with a fresh grin.

"Have you got a rope, Jack Sheppard?"

"Never mind the logs; they don't belong to me, and I don't care for 'em. Can't you lend me this boat a little spell? I want to git some saxifax over there for my mother."

"I can't spare her now; I have to go over to Westport after my sister," answered Paul.

"Be you, though? I guess I'll go over with you," said Tom, with refreshing confidence.

"I thought you were going to get some sassafras for your mother," added Paul, who had by this time come to the conclusion that his visitor was as queer as the craft upon which he had come.

"I guess I'll git it another time. I want to go over to Westport to see a feller I know there."

"But I can't bring you back, for my sister is coming with me, and the boat won't carry more than two," answered Paul, supposing this would end the matter.

"All right; I will stay with the feller I know all night," replied the accommodating stranger. "I'll help you row the boat over."

Paul was quite willing to have the queer fellow go with him, even if he had to row all the way himself, for the flatboat worked better with two in her than with one. Without a passenger she was too deep in the water forward, and dug her nose into the wave. He had not the least idea who his visitor was, but did not believe he had given his right name. If he had known him, he would have given him the cold shoulder at once.

"I am not ready to go yet; I have to fix up the boat a little," continued Paul, as he laid the oars on the beach.

"How long before you can go? I don't want to wait all day, Bristol Brick," added Tom.

"You need n't wait one minute if you don't wish to," replied Paul, who wondered in what school of politeness his involuntary companion had been brought up. "I am going to wash out the boat and let her dry a little before I go."

" What's the use of washing her out? She is clean enough for any feller to eat his dinner in," growled Tom.

" Perhaps she is if the fellow's a pig," said Paul, as he hauled the boat up on the beach, nearly upsetting the Topover in the act.

" Mind out! What are you doing? Do you mean to spill me into the drink?" demanded the saucy visitor.

" If you don't get out of the boat, you will be likely to get a ducking," added Paul, as he took up a pail and a broom he had brought from the house before.

Tom looked at the owner of the Dragon ; in fact, he eyed him from head to foot. Tom was a pugilist, or he thought he was. He was a fighting character, and possibly he was thinking whether he could whip the son of toil, whose story had been told all over Genverres as well as on the other side of the lake.

The Chesterfields had patronized and encouraged Tom Topover the season before, because they believed that he and his vicious companions could assist them in their encounters with the Beech Hillers.

During the present season the students of the two schools had not come in collision, for the reason that Colonel Buckmill had forbidden his pupils to meddle with their old enemies. Since the removal of the cottage from Sandy Point the Chesterfield barges had been down to Porter's Bay, where Tom had met Walker Billcord. The Topover and his gang had fully discussed the attempt to capture Paul Bristol, and were familiar with all the particulars.

Tom wished the magnate had offered *him* twenty-five dollars for capturing the fellow, and handing him over to the major at Sandy Point. This he said to Walk. He would be willing to do the job for ten dollars. He did n't think it was a great undertaking.

"You had better try it on, then," added Walk, morosely, for he hated Paul not a particle less than when the son of toil had laid him out on the point. "He will knock you out every time, without half trying."

"I should like to see the feller that can do that!" exclaimed Tom, clenching his dirty fists.

"Have n't you seen Paul Bristol?" asked Walk.

"I saw him on the other side of the pond; I never see him close to. But I ain't afear'd on him. I 'll bet I can lick him so he won't know whether it 's Sunday or Thanksgiving," blustered Tom. "Will your old man give a feller anything now for ketchin' him?" inquired Tom, looking anxiously into the face of Walk.

"He don't make any offers for him now," added the son cautiously. "He don't want to get into any more rows about the fellow." .

"Oh, he don't?" muttered Tom, evidently greatly disappointed. "If your old man will only do the handsome thing, I 'll scrape this Bristol Brick till there ain't nothin' left on him."

Walk Billcord looked at the ugly customer at his side, and his thirst for vengeance stirred all the bad blood in his veins. He had plenty of spending money, and he could even afford to give ten dollars himself for satisfactory vengeance. Nim Splugger and Kidd Digfield, as rough specimens as the Topover himself, would assist Tom. But Walk's father had just been discharged from confinement, and there was great risk in making the trade suggested.

"You would be prosecuted if you did anything,"

suggested Walk. "Then it would come out that I had a hand in the business."

"Not a bit on 't!" exclaimed Tom, very positively. "Jest as quick as I git the ten dollars, nobody won't see nothin' more of me within a hund'ed miles of Lake Champlain."

"What do you mean by that, Tom?" asked Walk curiously.

"I'm go'n to run away. My old man is so hard on me that I can't stand it no longer. I'm go 'n' to New York to ship in a pirate vessel. I sha n't be caught nohow."

"I would give ten dollars quick enough to see Paul tied to a tree and lathered with a cowhide for twenty minutes or half an hour; but I don't make any offers, and I won't hire any fellow to do such a thing," added Walk, as he considered the appalling risk.

"I cal'late I know jest what you mean, and you don't make me no offers. You don't promise to give me no money," protested Tom.

"What are you jawing about, Walk?" asked Ham Jackson, coming up at this moment.

"Tom Topover offers to catch Paul Bristol, and give him a lathering that will keep him on his bed a month, for ten dollars; but I won't do anything of the sort. I don't offer him a cent. I won't give him a penny if he kills the rascal," said Walk, with as much earnestness as though he meant every word he said.

"That's jest how it is. He won't give me nothin', and says he won't," added Tom.

" If you choose to larrup him on your own account, it is none of my business," continued Walk.

" Of course it is n't," Ham Jackson chimed in. "I should like to pay that fellow off for the few cracks he gave me, but they go to law on this side, and it is n't safe."

"Of course I can lick him if I want to, and 't ain't nobody's business," added Tom, who thought he was very cunning. "I guess I understand you, and you understand me. About next Saturday night at Sandy Point, say."

The coxswain's call summoned them to the boat, and they parted from Tom Topover. The latter believed he had made a square bargain with Walk Billcord, and ten dollars would take him to New York and pay his way till he could ship in a "pirate vessel." He meant a pilot-boat, for he had heard some one talking about one of these brisk little schooners a few days before.

Since that interview Tom had watched the school

grounds all the time. Paul lived on the point, and he could catch him alone there some evening. He had built the queer craft for use in his great enterprise. He had seen the Sylph go down the river in the morning, and he intended to put his scheme in operation that evening. Paul often sat on the rocks about dark, and the opportunity would not be wanting.

While he was nailing the logs together on the other side of the creek, a little way up, he saw Paul in his flatboat. Then it seemed to him that the son of toil was as good as bagged. He was absolutely sure he could handle him, in spite of the experience of the kid-glove chaps on the other side. But Tom was cunning in his own estimation. Paul was going to Westport, and it was safer to do the job near Sandy Point than on the school premises.

He could hardly help bullying, but he refrained as soon as he thought what he was doing ; and half an hour later he embarked in the flatboat with his victim.

CHAPTER XXVII.

THE BRILLIANT STRATEGY OF THE BRUISER.

THAT cunning Tom Topover was actually behaving himself in something like a civilized manner, in his desire not to alarm his victim. Just now he was engaged in a strategetic enterprise, and found it necessary to display only the fur side of his nature, though even that was very like the bristles of a pig. He did his best, which was not saying much, to keep on the right side of his intended victim. But Paul was a good-natured fellow, and it was an easy matter to conciliate him.

The son of toil rowed down the river, and crossing the shoal water of Field's Bay, took a straight course for his destination. Tom sat at the stern, and did not seem to be as much inclined to talk as he had been immediately after the wreck of the queer craft. In fact, he was turning over in his mind sundry cunning propositions, to accomplish the purpose for which he had embarked in the present venture.

It was a good six-mile pull to Westport, but Paul was used to the Dragon, and she went ahead without much effort on his part. The lake was as smooth as glass, and the rower wondered that it could ever be as rough as it had been the day the Silver Moon was so nearly wrecked. Though he was as tough as an oak knot, and had not yet become tired, he thought it was about time for the stranger in the stern to begin to do his share of the pulling, for the boat was now about half way to Westport. Sandy Point was half a mile ahead, and Paul mentioned the fact as a hint that his companion had better take the oars.

"They say you used to live there, Bristol Brick," said Tom in reply, and without taking the hint, which was altogether too indefinite for one with a skin so thick and dirty as the bruiser had.

"I lived there two years," replied Paul indifferently.

"Sho'! You don't say so!" exclaimed Tom, albeit there was nothing very astonishing in the statement. "They say the Beech Hill fellers kerried the house you lived in over to Hornet P'int one night, and left everything jest as though there never had n't been no house there."

"That's all very true. Major Billcord warned

my mother to move the cottage within twenty-four hours, and told her he should pitch it into the lake if she did n't do it," added Paul; and the stranger seemed to be the only person in Addison County who did n't know all the particulars of the affair.

"I guess the major was a little struck up when he found it had scooted," said Tom, with a cheerful grin, as he looked ahead at the point where he had suggested a meeting of the oarsman with Walk Billcord on the evening of that day.

"He was very much astonished, and so were the students of the institute, to whom he had promised a pile of fun in tipping the cottage and all that it contained into the lake."

"I don't see how the Beech Hill fellers could move the house. I don't believe they did it," added Tom, shaking his head.

Paul explained how the job had been done, and assured his companion he had seen the whole work himself. Tom insisted on being incredulous, for just then he believed he was particularly cunning.

"I never went ashore at Sandy Point, Bristol Brick, and I should like to see how the land lays there," suggested Tom, with one of his cheerful grins, exaggerated for the occasion.

" You can see the whole of the shore from here,"
replied Paul, turning around and pointing out the
locality of the cottage.

" But I want to see the place, and 't won't take
two minutes for me to run up to where the house
was," Tom insisted. " Then I will row the rest
of the way over to Westport, and nobody won't
git hurt none."

Paul had started more than an hour earlier than
he had intended, and would reach his destination
before Lily had finished her day's work. Besides,
he had a kind of affection for the place where he
had lived two years. Just then it flashed upon
his mind that he had never visited the hollow tree
which had done duty as a safe for the two dollars
and the gold rings belonging to his mother.

In the excitement of his last visit to the point at
the invitation of the magnate, he had forgotten all
about the treasure. His mother had spoken of it
often, but Paul had no doubt it was safe in its hid-
ing-place, for the money and rings had been put at
a tin box.

His mother had spoken of it, and so had he, at
the time of it, but latterly it seemed to have passed
out of the memory of all the family. Paul pulled
to the shore as soon as the treasure came into his

mind, and he wondered that he had not thought of it before. His mother had plenty of money now, and that seemed to be the reason it had been forgotten.

When Paul swung the boat around, and headed it for the point, he took a look down the lake. Over in the direction of Button Bay he saw a steam yacht. There were several such craft on the lake, though all or nearly all of them were kept farther down. The yacht looked exactly like the Sylph, and he had no doubt it was she.

"I wonder what the Sylph is doing over there," said he, continuing to pull for the beach before him. "She went down to Port Henry towing the gundalow with a cargo of stone."

"I guess the fellers are taking a little turn in her while the men are unloadin' the stone," suggested Tom, who was not at all pleased to find the Beech Hill steamer in this part of the lake.

"She is headed this way, and perhaps she is going back to Beech Hill after something that was forgotten," added Paul, as the Dragon struck the sand on the beach.

Paul took the painter in his hand and stepped ashore. He paused a moment to take another look at the Sylph. She was coming up from Button

Bay on the east shore of the lake, and this course would carry her within a mile of Sandy Point. It was now four o'clock in the afternoon, and the steamer, with her heavy tow, must have reached Port Henry by eleven at the latest. If anything had been forgotten, they must have found it out earlier in the day.

While he was looking at the steam yacht and wondering what she was doing in this part of the lake without her tow, she suddenly changed her course and stood over in the direction of Westport. This settled it that nothing had been forgotten, for she was not going back to Beech Hill. She was sailing very fast, and seemed to be shaken by the effort of her engine. They were certainly driving her at a very unusual speed.

Tom Topover was walking about the point, apparently engaged in a very minute inspection of the locality. Paul saw him looking at the former site of the cottage, and then he disappeared in the woods. The owner of the Dragon drew his boat a little farther up on the beach, but he continued to watch the movements of the steamer; and he was so absorbed in the effort to fathom her strange behavior that he was in danger of again forgetting the treasure in the tin box.

From his position on the point Paul could see
the steamboat wharf at Westport, or, rather, he
could see where it was, for it was over two miles
distant.   But the steam yacht did not go to it; and
for a short time she disappeared from his view
behind the trees on the lower arm of the point.
But he knew she must come in sight again soon,
for there was no landing-place above the wharf,
and the water was shoal.

In a few minutes she did reappear, and now she
was close inshore, following the southern trend of
the bay.   She had reduced her speed somewhat, but
she was still sailing faster than her standard rate.
Paul watched her till she reached Barber's Point,
behind which she again went out of sight.   He
could make nothing of her erratic movements, and
he was forced to the conclusion that Tom was
right, and that the fellows were taking a little
turn in her while waiting for the cargo of the
gundalow to be discharged, or for the iron shoe for
the keel of the Lily.

By this time Tom Topover appeared to have
completed his survey of the locality of the cottage,
and joined Paul on the beach.   The cunning fellow
seemed to be somewhat uneasy and excited, though
his companion was too much absorbed in the wonder of the steam yacht to notice it.

"Be you about ready to start on?" asked Tom, after he had looked about him for a few minutes. "I guess I've seen the whole thing now."

"I can't make out what the Sylph is doing," said Paul, still perplexed by the problem, though there was n't the least reason why he should bother his head at all about her strange movements; but, like the average boy of intelligence, he desired to know what everything meant.

"She's only cruising about for the fun on 't," grinned Tom. "I guess I don't want to stop no longer."

That cunning reprobate had arranged his plan of operations. In the darkness of the woods he had examined the tarred spun-yarn which filled one of his trousers pockets. He had taken it from a new building on the back road, where it had been used to secure bundles of laths. He had coiled up the single lengths in such a way that they would be ready for use when wanted. With these he intended to bind his victim hand and foot, and then tie him to a sapling, which he had selected for the purpose, in the woods back of the cottage site, where the prisoner could not be seen or heard from the lake.

He had promised to row the Dragon from Sandy

Point to Westport; and it was with a purpose that he had proposed to do so. Paul was to sit in the stern, and would have to get into the boat first. Tom would be close behind him, and when he took the first step, he would seize him by the throat, throw him down on the beach, and lie down on him. With the spun-yarn in his pocket he could easily secure his hands behind him. He had picked up a stout stick in the woods, which he dropped carelessly on the shore, where it would be available in case of need.

Tom had no doubt whatever of his ability to carry out this nicely-arranged programme. Paul was a stout fellow, and events at the point and elsewhere proved that he had plenty of pluck, and that he hit hard. But if he took him behind, what could Paul do? What could any fellow do, under such unfavorable circumstances? The blunder of the six ruffians, in Tom's estimation, was in attacking him in front instead of in the rear.

The cunning bruiser was ready to execute the plan his busy brain had contrived, and he was a little nervous and uneasy, as before noted. He did not take the least interest in the movements of the steamer, though he was rather pleased to find Paul so much absorbed in anything that kept his mind occupied.

"You git in fust, as I'm go'n to row the rest of the way," said Tom, as he took the oars from the boat, the blades of which were projecting over the bow.

"I am not quite ready to go yet; I have to go over after something I left in the hollow of a tree," replied Paul, as he turned away from the boat.

"In the holler of a tree!" exclaimed Tom.

"That's what I said," added Paul. "It is a tin box containing a little money and a couple of gold rings. It won't take me long to get them."

"How fur off is it?" asked the bruiser, much interested when he heard there was money in the box, for he was sure to get it.

Paul said it was up in the hollow, and started off.

CHAPTER XXVIII.

A HARD BATTLE AT SANDY POINT.

THE brilliant strategy of the bruiser was not affected by the visit of Paul Bristol to the hollow tree. On the contrary, it improved its prospects of success. The intimation that there was even a little money in the tin box was encouraging, for it would add something to the ten dollars he was to receive on account of his blind bargain with Walk Billcord.

As soon as Paul disappeared in the woods, Tom took the boathook from the Dragon, the oars being already in his possession, and hastened off in the direction to which he had carried his previous survey of the locality. He had found a low place beyond the site of the cottage, where a rotten log lay on the ground. Beneath this decayed wood he deposited the oars and boathook. Pulling off enough of the punky wood to cover the articles, he returned to the boat with a rapid step.

He had been absent but a few minutes, and

Paul had not yet appeared with the tin box. If the cunning strategist had been asked why he concealed the oars and the boathook, very likely he would have replied that he had done so to prevent the possibility of an escape on the part of his victim. But Tom was a cunning fellow, and this was by no means his object. If he failed to accomplish his purpose in the first onslaught, there is not the slightest doubt that he would have been entirely willing that his intended victim should escape, and even be glad to have him do so, even if he had been left to find his way on foot from the point.

The three implements which constituted the furniture of the Dragon might be dangerous weapons in the hands of a resolute fellow like Paul Bristol. He had secured a club for himself, and picking it up, he was plying it as a cane and plaything, in order to avert any suspicion as to its probable use.

Paul soon returned with the tin box in his hand. It was an old mustard can, and it was not a convenient thing to have in his pocket, and was of no value. He took the rings and money from it, and put them into his pocket, throwing away the can.

"How much money have you got, Bristol Brick?" asked Tom, with his usual grin.

"Only a two-dollar bill, and that belongs to my mother," replied Paul, who did not know his companion, and would not have been afraid of losing the money if he had.

"Ain't you goin' to spend it down to Westport, and treat a feller that helps you row the boat?" asked Tom, with a mighty grin.

"Of course not; I don't spend my mother's money for anything, without her orders to do so," replied Paul.

"What odds does it make?"

"It makes a good deal of odds to me, for I don't spend what don't belong to me. By the big wooden spoon! There comes the Sylph again, and without the gundalow in tow. She is going as though she were running a race too. It looks to me just as though Captain Dory Dornwood had gone crazy, and I shouldn't wonder if Captain Gildrock hauled him over the coals for it."

"Is the cap'n aboard?" inquired Tom.

"No; he went to Burlington this morning in the cars."

"Then he won't know nothin' about it."

"Some one will be sure to tell him; but the officers have to keep a log, and put down the speed every hour. I am afraid Dory will catch it, for that steamer is running fifteen knots an hour. They say she can do it, but the fellows are not allowed to do more than twelve."

"All right; but I guess we can watch her from the boat just as well as we can here," suggested the strategist, placing himself close behind his victim, and bracing up for the effort he was to make.

"If she comes near us, we can hail her, and find out what she is doing, for I should like to know," added Paul, as he stepped down to the beach, in the direction of the bow of the boat; but his eyes were fixed all the time on the steamer, which was certainly going like the Flying Dutchman.

"Jump in, Bristol Brick, and I will take the oars."

Paul had reached the stem of the Dragon by this time, still closely observing the steamer. He was just beginning to wonder if she would not blow up under such a tremendous pressure of steam as she appeared to have on, when the arms of the cunning strategist encircled his neck, and his right knee was applied to the small of his back. He had raised one foot to step into the boat, but he had no chance to bring it down, for he went over backwards on the beach.

The bruiser had the club in his hand when he passed his arm around the neck of his victim. In the suddenness of the attack Tom Topover had it all his own way, as he had intended to have it. As he drew his prisoner back, he threw him over so that he fell on his face, and Tom came down on top of him. He hugged him with all his might. Dropping the stick, he fixed his grip on the throat of Paul, and began to jam down upon him with his knees.

But Paul soon came to a realizing sense of his situation, and he was not at all inclined to submit to the sharp discipline of his companion. He began to struggle with all the energy of desperation. His hands were at liberty, and, reaching down with them, he succeeded in getting hold of the legs of his assailant. He immediately put a stop to the action of the assailant's knees, and then, with a mighty effort, rolled over so that Tom was under him, though Paul was still wrong side up.

With the weight of his victim upon him, Tom could no longer kick or use his knees, and Paul's hands were relieved for other duty. He brought them up and got hold of Tom's hair, getting two fistfuls of it, for the bruiser did not wear a fighting cut just then. He pulled with all his strength, increased by his desperation. At the same time the struggle with the other parts of the body continued, Tom's hair was coming out by the roots, and the intense pain caused him to yield a little of his hold at the prisoner's neck.

Paul felt his advantage, and, seizing the hands of his foe, dragged them from his throat. This enabled him to turn over in part so that he could use his fists. He did not wait for any preliminaries, but rained his blows upon the head of his assailant in the agony of his desperation. Tom could no more stand this treatment than he could have endured the pounding of a trip-hammer. He begged for mercy, and Paul let him up.

Neither of them could speak, and Tom's dirty face was covered with blood. Both were gasping for breath, and an involuntary truce prevailed. Paul had received no blows in the face, though his throat was considerably lacerated by the nails of his cowardly enemy. Tom was now in a position to understand the reason why the six ruffians had been so badly used before they succeeded in making a prisoner of Paul. It seemed to make no difference in the end whether the attack was made in the front or the rear. Possibly, the brilliant strategist was willing to believe that he had made a mistake in the quality and quantity of his intended victim.

A couple of minutes were enough to enable the combatants to recover their breath. Neither of them said a word, but Tom suddenly made a spring at Paul, this time with clenched fists. But the latter had been looking for something of this kind, and he easily parried the blows aimed at him, and then upset the bruiser with a heavy blow between the eyes. Paul realized that he could do this sort of thing·till the sun went down, but he was tired of it.

Tom lay still for a minute or so after his fall, for his ideas were doubtless greatly confused. Paul looked at him; and as he did so he saw one of the coils of rope-yarn sticking out of his trousers pocket. He seized it at once, and, turning his assailant over, tied his hands behind him, and then secured his arms at the elbows. As Tom came to a realizing sense of his defeat, he began to resist, but the bruiser was about played out, and Paul dragged him to a tree and made him fast.

"You don't fight fair, Bristol Brick," said he, rather feebly, and he made a weak attempt to break from his bonds.

"I don't mean to fight fair with such fellows as

you are," replied Paul, looking with disgust at his prisoner. "I suppose you do, though, and that's the reason you pitched into me when my back was turned. You may call it fair to jump on a fellow's back and pull him down."

"But 't ain't fair for you to tie me afore we have done," groaned Tom. "That's mean, and Tom Topover never lets up on a feller that don't fight fair."

"Oh, then, you are Tom Topover, are you?" exclaimed Paul, looking over his victim from head to foot. "Why did n't you tell me who you were when I asked you?"

"I was afear'd you'd run away if I told you who I was," answered the bruiser, who was likely to be a bully to the end.

"I don't believe I should have run away," added Paul, with a smile on his face. "I don't run away from such carrion as you are."

"You have n't seen the end of this thing yet. I can lick you in fair fight any time," blustered Tom, as he began to regain his strength.

"Will you do it now if I let you loose?" demanded Paul sharply.

"I don't feel very well to-day," replied Tom, after some hesitation. "I ain't in fightin' trim no-

·how, and that's the reason I got the worst on't so fur."

"What did you pitch into me for if you are not in good condition?" demanded Paul, who was good-natured enough by this time to smile.

"I did n't think you was so much of a feller, and I had to do what I did to-day," muttered Tom.

"Why to-day?" demanded Paul.

"Well, I agreed to do it."

"Whom did you agree with?" continued Paul, picking up the stick the mighty strategist had brought from the woods.

"It don't make no difference," whined Tom, evidently startled when he saw the weapon in the hand of his conqueror.

"Yes, it does make all the difference in the world; and if you don't tell me in two seconds, I will take it out of your hide!" exclaimed the son of toil, demonstrating violently with the stick.

"I did n't agree to do it, but Walk Billcord was to give me ten dollars for the job. He did n't say he would, but we understood one another," answered Tom, in mortal terror.

"That's all I want to know," added Paul, as he walked towards the boat.

He looked into the Dragon, but did not see the oars. He searched all about the beach without being able to find them. While he was thus engaged, the steamer came within a few feet of the shore. He concluded that the absence of the oars was a part of the cunning strategist's plan; and he was about to return to the tree where Tom was tied, when the steamer rang one bell, followed by two. This meant stop and back her.

Paul picked up the stick he had brought to the water side, and, without looking particularly at the Sylph, he pushed off the boat, and then gave it a hard shove with the short pole. The impetus carried the Dragon to the side of the steam yacht, and he sprang on board of her with the painter in his hand.

CHAPTER XXIX.

THE ENGINEER OF THE UNDINE.

PAUL BRISTOL was somewhat excited after his tough conflict with the strategetical bruiser. He was not a little startled to find that the Billcords were still trying to punish him for defending his sister from insult. Captain Gildrock was his friend and his mother's friend, and he was unwilling to do anything more with the pestilent bully without his advice and direction. He was confident that the exhibition of the stout stick would induce Tom to tell him where the oars were; but as the steamer was close to the shore, he preferred to take counsel before he acted any further.

At first he forgot that the principal was not on board of the Sylph, but it came to his mind before he reached the bulwarks. But Dory was certainly on board, and he could advise and assist him. Passing the painter over a stanchion, he leaped over the rail. Then it struck him as a little strange that he saw none of the large ship's company that had manned her when she left the wharf in Beechwater that morning.

A man who was an entire stranger to him stood on the forecastle, but not a single Beech Hiller was to be seen. He looked up at the windows of the pilot-house, where he expected to see the face of Oscar Chester and the second pilot; but another stranger stood at the wheel.

"Cast off that boat!" called the man at the wheel to the one on the forecastle.

Before Paul could interfere the deck hand had detached the painter from the stanchion and dropped it into the water. At the same moment two bells rang, and the steamer backed away from the point.

"What did you do that for?" demanded Paul of the deck hand.

"I have to obey my orders," replied the man.

The son of toil looked at him and wondered who he was, for he had never seen his face before. He went to the bow and saw the Dragon, fifty feet from him by this time, and the steamer still backing. He had been sure of obtaining good advice and strong support from his friends on board, but he could not even find a person that he knew.

He walked aft, and looked into the engine-room. There was a man there, but he was bending over the machinery, and he did not see his face, but he appeared to be a stranger like all the others he had seen. He continued his walk to the door of the after cabin, but not a single Beech Hiller could he find. It looked to him as though, if the thing had been possible, the Sylph had been captured by an enemy, who were then in full possession of her. ·

Paul returned to the forecastle, and again looked up at the windows of the pilot-house. The man at the wheel appeared to be talking to some person or persons behind him, who were not in sight. At this moment the engine stopped again, and the steamer was at rest on the smooth water. Paul was confident that the persons in the pilot-house had seen the whole or a portion of the hard battle at Sandy Point, for the tree where he had secured Tom Topover was in plain sight from the lake.

The deck hand seemed to take no notice of him, though he could not help seeing him, and observing all his movements. As the steamer had stopped her propeller, and run close in to the shore, she must be there for a purpose.

The last he had seen of her before the bruiser opened the fight, she was running with tremendous speed down the lake. After that she slowed down, and headed for the point, for he had obtained an occasional glimpse of her even in the heat of the struggle with the strategist.

"Is Captain Dory Dornwood on board, sir?" asked Paul, very respectfully, of the deck hand, who had walked forward to look out.

"I don't know him," replied the man shortly, but civilly enough.

"Are none of the Beech Hill fellows on board?"

"I don't know the Beech Hill fellows," answered the man.

Paul was utterly bewildered. He looked up at the pilot-house once more to find a solution of the mystery if he could. The stranger still stood at the wheel, and was still talking with some one not in sight. Just then it occurred to Paul that there was something wanting in the appearance of the pilot-house. In the station bill he belonged on the forecastle of the Sylph, and was more familiar with this part of her than with any other. He studied the situation for some time before he could determine what was wanting to complete the usual appearance of the steamer. At last he was able to supply the deficiency. On the front and on each side of the pilot house was a sign on which was painted the name of the craft. They were not there; and if the strangers had captured the Sylph they had removed these signs. There was nothing in sight to indicate that the vessel was the Beech Hill steam yacht.

Paul looked around him on the forecastle, and some other familiar objects were missing. Suddenly it flashed into his mind that this was not the Sylph after all; but the absurdity of his making a mistake in the identity of the steamer which he was accustomed to see at the wharf in Beechwater every day, and in which he had made so many trips to Westport and elsewhere, was so apparent to him that he instantly rejected the idea.

To his mind, in spite of the absence of the signs on the pilot-house, and other familiar objects, the steamer was the Sylph. The mystery of her being in possession of other persons than the Beech Hillers seemed to thicken upon him. She had taken position not fifty feet from the water side, and there she lay. Paul wondered what she was waiting for, and why she did not do something. If any person on board was to be landed at the point, it was about time to lower one of the quarter boats, which hung on davits abaft the engine. But nothing was done, and no one said anything; and Paul was getting desperate.

There was a mystery about the steamer, which, in spite of the good order which prevailed on board of her when her regular ship's company were on duty, presented a very lively aspect. Paul was not patient in the presence of mysteries which concerned him, as in the present instance, for since the setting adrift of the Dragon he was practically a prisoner on board of her. He decided to solve the problem of the strangeness of things on the deck, and for this purpose he went aft to the ladder on the port side which led to the hurricane deck. He was determined to have a pow-wow with the pilot, and to ascertain who were the modest persons that concealed themselves in the back part of his quarters.

He reached the deck on which the pilot-house stood, without impediment, and walked to the door. It was locked, which was not usual on board of the Sylph. He passed on to the side window, where the man at the wheel suddenly confronted him. He had seen this man before, but he had no acquaintance with him.

"Will you be kind enough to tell me what steamer this is?" Paul began, in his efforts to solve the mystery.

"The Undine, of Westport," replied the pilot, for such he undoubtedly was.

"I never heard of her before," added Paul, overwhelmed to find that she was not the Sylph.

"As she came into the lake for the first time this morning, you were not in the way of hearing of her," answered the man, rather stiffly.

"But she looks exactly like the Sylph, which belongs to the Beech Hill Industrial School," said Paul, trying to get a sight of the persons on the sofa of the room.

But the pilot kept himself directly in front of him, and he was unable to gratify his curiosity.

"The builder has sent out at least half a dozen steam yachts of the hundred-feet order which are so near like this one that you could not tell the difference in them," added the pilot in answer to his remark.

"If this is not the Sylph, I have no business on board of her," continued Paul. "I shall be very much obliged to you if you will put me on shore, or pick up my boat for me, though I am very sorry to trouble you."

"You were not invited on board, and you must look out for yourself."

"Why did you cast off the painter of my boat?" asked Paul, not pleased with the situation.

"I obey my orders."

"Who gave the order, if you please?"

"The owner," replied the pilot. "Something was going on ashore there just before we came over here. It looked like a very hard fight between two fellows."

"It was a hard fight," answered Paul.

"And you were one of the fellows in it?"

"I was; and the other fellow is tied to a tree on shore," replied Paul, pointing in the direction of the tree, which he could see from his position on the hurricane deck.

"What was it all about? Speak up a little louder, for I am rather deaf," added the pilot, as he glanced behind him. "Who was the other fellow?" And the last question seemed to be prompted by the person on the sofa.

"It was Tom Topover," answered Paul; and in answer to questions put by the pilot, he told the whole story of his affair that day with the brilliant strategist, from the time he had appeared in the creek on his queer-looking craft.

The pilot occasionally told him to speak louder, and at last he concluded that he was giving the narrative for the benefit of the concealed listeners.

"But what made Tom Topover attack you?" asked the man at the wheel.

"He was hired to do it by Major Billcord's son, Walk Billcord," replied Paul bluntly.

"Do you mean to say that my son hired that rough to attack you?" demanded the magnate of Westport, suddenly rushing to the door of the pilot-house, and throwing it wide open. Close behind him was Walk himself.

"I didn't know you were here!" exclaimed Paul, starting back with astonishment when he saw the major; and he had not had the remotest suspicion that he was the owner of the steam yacht, for the pilot had prevented him from asking who owned the craft.

"No matter if you didn't know it," replied the major angrily. "I asked you a question. Answer it!"

"Tom Topover said he was to get ten dollars from your son for doing the job. He didn't say Mr. Walker agreed to give him the money, but there was an understanding between them to this effect," replied Paul.

"Tom Topover is a liar!" exclaimed Walk.

"He was to do the job to-day; and you seem to be here at Sandy Point to attend to the prisoner if Tom got him," added Paul.

Major Billcord was not in the habit of controlling his wrath, and he made a spring at the son of toil; but Paul beat a hasty retreat, for he dreaded another encounter with the magnate. He went aft and descended to the main deck; but he soon discovered that he was not pursued. He heard two bells in the engine-room, and the Undine began to back. Paul came to a halt under the starboard quarter boat, and devoted himself to an examination of the falls by which it was secured to the davits.

While he was thus engaged, one bell struck in the engine-room, followed by another, and the Undine went ahead. A moment later the jingle bell rattled, and the craft began to go at full speed. Paul heard steps on the hurricane deck above him, and he concluded that the major and Walk were after him. He walked astern to the doors of the main cabin. They were open, and he decided to retreat into this apartment if he was pursued.

"Engineer!" called Major Billcord.

"On deck, sir," replied the man in charge of the engine, as he stepped out of his room.

The sound of the engineer's voice was a familiar one, and it startled the son of toil as much as the sound of an earthquake would at that moment.

"Keep an eye on that boy down on the main deck, and don't let him touch the boats," continued Major Billcord, who suspected the purpose of the object of his hatred. "Don't let him escape on any account, for I shall have a reckoning with him before we part."

This looked like a threat, and Paul realized that he was in the hands of the enemy. In spite of his imprisonment, the magnate intended to punish him for what he had done at the point, and the poor fellow began to be very much discouraged.

"I will see to him," replied the engineer.

The engine of the Undine was working at a moderate speed, and the engineer walked aft to get a view of his prisoner. Paul looked at him as he approached, for the sound of his voice had prepared him for an early meeting. He thought no more of getting away in the boat. He gazed with all his eyes at the man walking towards him.

"Why, father!" exclaimed he, rushing upon him with extended hand.

"Why, Paul, my son! Is it possible that it is you?" cried the engineer, grasping the extended hand. "But come into the engine-room."

Mr. Bristol led the way, still holding his son's hand. They had scarcely entered the apartment before there was a whistle at the speaking-tube.

"Take that boy into the engine-room, and don't let him get away," said the magnate through the tube.

"All right; I have him here," replied the engineer. "What does all this mean, Paul?" asked the astonished father, turning to his son.

Paul related all the events in the family history since the assault upon Lily at the point; and the returned wanderer fully understood the feud between Paul and the magnate. His blood boiled at the insult to his daughter, and the persecution to which his son had been subjected. He had put his hand on the wheel to shut off the steam, when Paul asked him where he had been for two years, and why he had not written to his family.

The engineer did not turn the wheel, for the wanting letters were an imputation upon him. He was not a scholar, but he had written a score of letters and had never had a reply to one of them. Before he left, something had been said between himself and his wife about her going to the home of an uncle in Iowa. He had invited them to visit him and take care of him, for he was a bachelor. He would support them, and they could do work enough to earn their living. They had expected to hear from him every day at the time Peter Bristol left home.

The father had no doubt they would go there, and had directed his letters after the first one to their new home. A few days after his departure for New York, where he hoped to find work, the letter came from the West to Mrs. Bristol, but it brought no hope. The writer had bought a ranch in Texas, had married, and could do nothing for the family of his brother. This clearly explained the miscarriage of the letters.

Peter Bristol had worked as a fireman on a railroad. When he got to New York he found a situation as an oiler on a steamer bound to Havana. In Cuba he soon secured a good situation to run an engine on a plantation. He saved his money, and did his best to find what had become of his family. At last it occurred to him to write to the postmaster of his brother's late residence in Iowa. Nothing was known of his family, his brother had gone to Texas, and a score of letters for his wife had gone to the dead-letter office.

Then he had written to a friend in Westport, and learned that his family were still at Sandy Point, and were very poor. When this last letter came, nearly two years after he had left home, he was filled with sorrow and anxiety. He wrote no more letters, but started for home with all the money he had saved. About the first person he met when he landed in New York was Wheeler, whom he had known as a pilot on Lake Champlain. He had been sent by Major Billcord to take his steamer, just purchased, up to the lake by the way of the Hudson and the canal. He wanted an engineer, and, after a deal of talk, employed Peter Bristol.

Wheeler had his doubts about the competency of Bristol. The magnate wanted a suitable engineer, and would give him good wages. He might object to a man who had been known on the lake as nothing but a boatman. Peter wanted the place, and had been running an engine for two years. Wheeler agreed to do what he could for him with the magnate; but he thought it best for him not to say who he was for the present. Time and the tropical sun had so changed him that he was not likely to recognize him if he was careful.

Peter Bristol had served as engineer on the way up, and Major Billcord and Walk had joined the vessel at Whitehall in the morning. The steamer was on trial, and the major wanted her run at her highest speed a part of the time. The magnate had hardly looked at the engineer, he was so interested in the machinery and the craft, and Bristol had had no trouble in concealing his identity so far. This was the story he told Paul, and repeated to his wife and Lily in the evening.

Paul had looked out at the door and saw that the Undine was near Westport. She did not go to the shore, but when she came about and headed down the lake again, Peter Bristol turned the wheel and shut off the steam. There was a ringing of the bell, and then a call through the tube.

"I shall run her no longer!" replied the engineer, emphatically, at the mouthpiece.

Major Billcord came below, followed by Walk. Mr. Bristol stated his position, and took no further pains to conceal his identity. The father spoke to him like a man, and insisted upon being landed at Westport with his son. The magnate was taken all aback. He could do nothing without an engineer, and he could not punish Paul in the presence of his father. The engineer would take the steamer up to the wharf, but in no other direction. The magnate had to yield, and father and son, both the Bristol and the Billcord, landed.

Lily was found, and she had a joyful meeting with her father. Bissell was very willing to loan his four-oar boat to convey them to Beech Hill. On the way they released Tom Topover, and, putting him into the Dragon, towed him back to Hornet Point. The happy re-union in the transplanted cottage need not be described.

CHAPTER XXX.

LAUNCHING THE BOAT.

THE Sylph, with the gundalow, did not arrive till it was nearly dark. The shoe was not done when the scow was ready to take it on board, and they had to wait for the workmen to drill the holes for the bolts. The ship's company had seen the Undine when she passed Port Henry, but no one there knew to whom she belonged, or anything whatever in regard to her. They saw that she was the counterpart of the Sylph, and knew that she was one of the celebrated class to which she belonged.

Some of the students thought there might be a chance for a race between her and the Beech Hill steamer; but Dory was sure enough that Captain Gildrock would not permit the Sylph to race with anything that went by steam.

The principal had returned from Burlington in the afternoon, and when he saw the four-oar boat, with the Dragon in tow, moving up to Hornet

Point, he walked over to the cottage. He was a spectator of the affecting interview between Mrs. Bristol and her husband, even before the Dragon was hauled up to the shore.

Tom Topover was very much battered in the conflict with Paul. He was sure of two very black eyes, and he could hardly walk when he was helped out of the flatboat. The principal thought he had been punished enough for the present; and as he seemed to be very humble, for him, he was allowed to limp home, after a strong admonition from the captain.

The principal had been so good a friend to the family, that Mrs. Bristol begged him to stay and hear her husband's story, and listen to the adventures of Paul since he left in the afternoon. The moving of the cottage had to be related by Paul. The prolonged corversation was interrupted only by the arrival of the Sylph. After the shoe was landed at the boat-shed, and the gundalow towed to the stone quarry, the students learned all about the new steamer, in which they were very much interested, though they were sorry to learn that Major Billcord was her owner.

On Monday afternoon, the shoe was bolted to the keel of the Lily, and the inside work, which

had been left unfinished for this job, was completed. The following Saturday was appointed
for the launch of the boat, for this day would
complete the school year of the institution. Invi-
tations had been sent to the gentlemen who had
served as examiners the preceding year, and on
Friday afternoon the Sylph, fully manned, and
dressed in gay colors, brought up Mr; Bridges, Mr.
Ritchie, and Mr. Plint. They were hospitably
entertained at the mansion.

In the forenoon a sort of public exhibition took
place in the great hall of the boat-house, which
delighted the spectators, and gave them a very
high idea of the progress of the students in the
mechanic arts, as well as in the book studies.
After this show, the visitors went through the
shops, and inspected the Lily as she stood on the
stocks. A brass band played a portion of the
time, and in the middle of the day a choice collation was served on the green.

About all the young ladies in Genverres, and
not a few from Burlington and Westport, were
present. Possibly there was some heavy flirting
done, for again the students in their uniform were
lions of the first order.

But the great event of the day was to be the

launch of the Lily. She had been fully prepared in the morning for the exciting occasion, and two jury-masts had been put up on board, and she was covered with flags and streamers. The boat was to move from the ways at four, and an hour before that time the students and the principal were not a little astonished to see the two Chesterfield barges pull into the Beechwater, and take positions near the farther side of the lake.

Captain Gildrock sent Mr. Bentnick, the principal instructor, to invite them on shore to partake of a collation. Colonel Buckmill sat in the stern-sheets of the Dasher, but he politely declined the invitation, with profuse thanks. His young gentlemen desired to see the launch, but he would not give the principal any trouble on such a busy day.

The captain was sorry for this refusal, but he seemed to insist that the hospitality of Beech Hill should not suffer in the estimation of the students from the other side, and he sent a boat loaded with ice-cream, cake, and lemonade to the unwonted visitors, which were accepted with more thanks.

At a few minutes before four, the principal and a small party, including not more than half a dozen of the students, went on board of the Lily. A little later, the gallant captain of the Sylph escorted Miss Lily Bristol to the deck of the boat. Her appearance was the occasion of the most tremendous applause on the part of the students and the crowd assembled on shore. It was observed that the Chesterfields joined in this demonstration, with a vigor which astonished their former foes.

Captain. Gildrock gave certain orders, which were followed by the sound of hammers as the hands knocked away the wedges. The principal raised his hand, which was followed by one discharge of a cannon. At this instant the hull began to move very slowly. Assisted by Captain Dornwood, Lily Bristol ascended to the heel of the bowsprit with a bottle in her hand.

What this bottle contained no one but the captain knew. According to tradition and custom, it ought to be filled with wine; but the principal was a very strong, practical temperance man. However, as the contents of the bottle were to be dashed into the lake, it did not much matter what they were.

The velocity of the moving hull increased as she descended the inclined plane; and as soon as she was under full headway, Lily broke the bottle over the bow of the schooner.

"I give to this vessel the name of Lily, and may she be prosperous on the element to which she belongs," said she.

Then the band struck up "Hail to the Chief," and all the students and everybody else yelled and applauded with all their might. The ladies waved their handkerchiefs, and a salvo of artillery followed. The Lily struck the water, and ploughed her way nearly to the other side of the lake, where she was brought up by the lines attached to her. She rested on the water as gracefully as a swan, and as soon as she was fairly afloat, another series of ringing cheers saluted her.

The Sylph, under the charge of the first officer, immediately fastened to her, and she was towed to her berth at the wharf, where she was to remain until the next school year began, in September. But the visitors were eager to examine her, and an arrangement was made by which all who desired could pass on board, make the circuit of her deck and then leave without causing an uncomfortable crowd. Through the cabin doors and the opening for the skylight they could see something of the cabin, while the fore-scuttle gave a partial view of the cook-room.

The young officers and crew of the Sylph told their friends they should see her after she was rigged, her sails bent on, and the cabin and cook-room were furnished. There could be no doubt, so far as the students were concerned, that the young ladies who gushed so prettily over the craft would be invited to sail in her.

With this great event ended the second school year of the Beech Hill Industrial School. The students were certainly satisfied with the experience they had had there, and, notwithstanding the sharpness of the discipline, they had only pleasant memories of the past. Those who had been there two years were well prepared to earn their own living. Though none of the machinists or carpenters could be called finished workmen, they were skilled enough to obtain moderate wages. It would require more years of study and practice to make them into first-class mechanics.

None of them had yet completed the course of instruction, though the needs of their parents compelled seven of them to leave the school and assist in supporting families. By this time the reputation of the school had been established, and there were applications for three times as many young men to work as engineers, carpenters, and machinists. Good places were secured for those who were obliged to leave.

Three of them were to run stationary engines, one was to work as a carpenter, and three more were to learn trades for which their education had fitted them to a considerable degree. The principal had given them a lecture on the subject of wages, in which he bluntly told them that they could not expect full wages, for they were not competent to earn them. They were not yet physically able to do the work of a man, and they were not competent to do all that would be required of them in their several trades and callings. They had learned a great deal, and had acquired considerable dexterity; but if they were judged by what they did not know, they would stand as weak vessels. No man ever learned out in his trade, and the time never came when there was nothing more to learn.

A certain very wise man, as men are measured, declared that he had only learned enough to realize what a fool he was. The principal told the graduates that one of their greatest perils was that of knowing too much. Modesty in regard to the measurement of their own skill and knowledge was essential to them. It was better that others should find out how much they knew rather than themselves.

On Monday morning the Sylph went up the lake with the examiners and others who were to spend their vacations at home. In the afternoon she went down the lake with those who were going in that direction, and the ship's company was considerably smaller when the steamer returned to Beech Hill.

Mr. Bristol went on the afternoon trip, for Corny Minkfield and John Brattle, the engineers, were to be left at Burlington. Mr. Jepson was privately instructed by the principal to test his qualifications. It appeared that while he had but little scientific knowledge, he was as competent to run an engine as the majority of those who were employed in this capacity.

"I confess, Mr. Bristol, that I am very much interested in your family, and I shall be glad to retain you at the school," said Captain Gildrock. "I find that Mr. Jepson's duty in connection with the running of the engines in the shops interferes with his usefulness as an instructor. I shall relieve him entirely of the laborious task he has hitherto performed in the most faithful manner, for he is too valuable as a teacher to have any of his time wasted. I shall appoint you as engineer of the shops, though you are to serve in the steamer when required."

"I thank you, sir, with all my heart," replied Mr. Bristol. "You have done so much for my family, that I already owe you a debt of gratitude I could never repay."

"What I have done has afforded me as much pleasure as it has the members of the family," added the principal.

"The moving of the cottage was the greatest and the most timely thing that ever was done. But, Captain Gildrock, I saved considerable money, for a poor man, and I should like to buy a lot of land for my wife's cottage, and put a cellar under it."

"How do you like its present location?" asked the captain.

"Paradise has no finer spot, sir."

"Then I will give you a deed of the lot on which the house stands, without any money, for it will be a great protection to my estate to have your family in just that locality."

Mr. Bristol was overwhelmed at this generous

offer, and he accepted it with a heart full of gratitude. During the vacation the house was raised
somewhat and a cellar put under it. The Topovers,
who had troubled the captain for years by their
incursions, ceased to come in by the road to the
stone-quarries.

Though it was vacation at Beech Hill, and only
a very few of the boys, who had no homes, remained
during the summer months, there were some lively
times there. The instructors were all gone, but
plenty of company came from the cities. Almost
every day there was some kind of an excursion,
and Mr. Bristol was available as engineer, so that
the Sylph was constantly in use.

Captain Gildrock had another idea come into his
fertile brain. The Lily was so great a success that
he decided to build a steam yacht about half the
length of the Sylph, and to have the engine constructed in the shops by the students. He found
they enjoyed their work more when they were
doing something which they could use when it
was completed. However, he did not say much
about it.

Dory Dornwood, though his uncle suggested a
trip to New York, Niagara, or Montreal, found
more pleasure in staying at home, strange as it

may seem. Paul made himself useful as fireman or deck hand on board of the steam yacht, or as a foremast hand in the Goldwing. Mrs. Bristol and Lily were almost always passengers when either craft made an excursion, and so were Mrs. Dornwood and Marian.

The young captain of the Sylph spent a good deal of time at the cottage on Hornet Point, possibly because he had conducted the enterprise of transporting the structure from Sandy Point to its present locality. The captain and Dory's mother laughed a great deal about his constant visits, but as he never called the young lady anything but "Miss Bristol," the visits were not regarded as dangerous for the present.

Major Billcord and Walk sailed the Undine all over the lake, but she was apt to give the Sylph a wide berth. The testimony of Tom Topover was not considered sufficient to convict Walk of bribing him to capture Paul. Certainly, Tom got the worst of it, and the magnate was content to let the punishment of the son of toil go by default.

The third school year opened as prosperously as either of its predecessors, and the next volume of the series will doubtless contain as many of the adventures of the students on the lake and elsewhere as the former ones, as well as complete the unfinished work on the Lily; and, of course, the reader will find everything "All Taut" about the schooner after they have finished "Rigging the Boat."

OLIVER OPTIC'S BOOKS.

ARMY AND NAVY STORIES.

Six Volumes. Illustrated. Per vol., $1.50.

1. **THE SOLDIER BOY;**
   Or, Tom Somers in the Army.

2. **THE SAILOR BOY;**
   Or, Jack Somers in the Navy.

3. **THE YOUNG·LIEUTENANT;**
   Or, Adventures of an Army Officer.

4. **THE YANKEE MIDDY;**
   Or, Adventures of a Navy Officer.

5. **FIGHTING JOE;**
   Or, The Fortunes of a Staff Officer.

6. **BRAVE OLD SALT;**
   Or, Life on the Quarter-Deck.

This series of six volumes recounts the adventures of two brothers, Tom and Jack Somers, one in the army, the other in the navy, in the great civil war. The romantic narratives of the fortunes and exploits of the brothers are thrilling in the extreme. Historical accuracy in the recital of the great events of that period is strictly followed, and the result is not only a library of entertaining volumes, but also the best history of the civil war for young people ever written.

YOUNG AMERICA ABROAD.

FIRST SERIES.

**A Library of Travel and Adventure in Foreign Lands. 16mo.
Illustrated by Nast, Stevens, Perkins, and others.
Per volume, $1.50.**

1. **OUTWARD BOUND**;
   Or, Young America Afloat.

2. **SHAMROCK AND THISTLE**;
   Or, Young America in Ireland and Scotland.

3. **RED CROSS**;
   Or. Young America in England and Wales.

4. **DIKES AND DITCHES**;
   Or, Young America in Holland and Belgium.

5. **PALACE AND COTTAGE**;
   Or, Young America in France and Switzerland.

6. **DOWN THE RHINE**;
   Or, Young America in Germany.

The story from its inception and through the twelve volumes (*see Second Series*), is a bewitching one, while the information imparted, concerning the countries of Europe and the isles of the sea, is not only correct in every particular, but is told in a captivating style. "Oliver Optic" will continue to be the boy's friend, and his pleasant books will continue to be read by thousands of American boys. What a fine holiday present either or both series of "Young America Abroad" would be for a young friend! It would make a little library highly prized by the recipient, and would not be an expensive one. — *Providence Press.*

YOUNG AMERICA ABROAD.

SECOND SERIES.

**A Library of Travel and Adventure in Foreign Lands. 16mo.
Illustrated by Nast, Stevens, Perkins, and others.
Per volume, $1.50.**

1. **UP THE BALTIC;**
   Or, Young America in Norway, Sweden, and
   Denmark.

2. **NORTHERN LANDS;**
   Or, Young America in Russia and Prussia.

3. **CROSS AND CRESCENT;**
   Or, Young America in Turkey and Greece.

4. **SUNNY SHORES;**
   Or, Young America in Italy and Austria.

5. **VINE AND OLIVE;**
   Or, Young America in Spain and Portugal.

6. **ISLES OF THE SEA;**
   Or, Young America Homeward Bound.

" Oliver Optic " is a *nom de plume* that is known and loved
by almost every boy of intelligence in the land. We have
seen a highly intellectual and world-weary man, a cynic whose
heart was somewhat imbittered by its large experience of
human nature, take up one of Oliver Optic's books and read
it at a sitting, neglecting his work in yielding to the fascina-
tion of the pages. When a mature and exceedingly well-
informed mind, long despoiled of all its freshness, can thus
find pleasure in a book for boys, no additional words of rec-
ommendation are needed. — *Sunday Times.*

WOODVILLE STORIES.

Uniform with Library for Young People. Six vols. 16mo. Illus
trated. Per vol., $1.25.

1. RICH AND HUMBLE;
Or, The Mission of Bertha Grant.

2. IN SCHOOL AND OUT;
Or, The Conquest of Richard Grant.

3. WATCH AND WAIT;
Or, The Young Fugitives.

4. WORK AND WIN;
Or, Noddy Newman on a Cruise.

5. HOPE AND HAVE;
Or, Fanny Grant among the Indians.

6. HASTE AND WASTE;
Or, The Young Pilot of Lake Champlain.

Though we are not so young as we once were, we relished these stories almost as much as the boys and girls for whom they were written. They were really refreshing even to us. There is much in them which is calculated to inspire a generous, healthy ambition, and to make distasteful all reading tending to stimulate base desires. — *Fitchburg Reveille.*

THE STARRY FLAG SERIES.

Six volumes. Illustrated. Per vol., $1.25.

1. **THE STARRY FLAG;**
   Or, The Young Fisherman of Cape Ann.

2. **BREAKING AWAY;**
   Or, The Fortunes of a Student.

3. **SEEK AND FIND;**
   Or, The Adventures of a Smart Boy.

4. **FREAKS OF FORTUNE;**
   Or, Half Round the World.

5. **MAKE OR BREAK;**
   Or, The Rich Man's Daughter.

6. **DOWN THE RIVER;**
   Or, Buck Bradford and the Tyrants.

Mr. Adams, the celebrated and popular writer, familiarly known as "Oliver Optic," seems to have inexhaustible funds for weaving together the virtues of life; and notwithstanding he has written scores of books, the same freshness and novelty runs through them all. Some people think the sensational element predominates. Perhaps it does. But a book for young people needs this; and so long as good sentiments are inculcated such books ought to be read. — *Pittsburg Ga-zette.*

THE ONWARD AND UPWARD
SERIES.

Complete in six volumes. Illustrated. In neat box.
Per volume, $1.25.

1. **FIELD AND FOREST;**
   Or, The Fortunes of a Farmer.

2. **PLANE AND PLANK;**
   Or, The Mishaps of a Mechanic.

3. **DESK AND DEBIT;**
   Or, The Catastrophes of a Clerk.

4. **CRINGLE AND CROSS-TREE;**
   Or, The Sea Swashes of a Sailor.

5. **BIVOUAC AND BATTLE;**
   Or, The Struggles of a Soldier.

6. **SEA AND SHORE;**
   Or, The Tramps of a Traveller.

Paul Farringford, the hero of these tales, is, like most of this author's heroes, a young man of high spirit, and of high aims and correct principles, appearing in the different volumes as a farmer, a captain, a bookkeeper, a soldier, a sailor, and a traveller. In all of them the hero meets with very exciting adventures, told in the graphic style for which the author is famous. — *Native.*

FAMOUS "BOAT-CLUB" SERIES.

Library for Young People. Six volumes, handsomely illustrated. Per volume, $1.25.

1. ## THE BOAT CLUB;
   Or, The Bunkers of Rippleton.

2. ## ALL ABOARD;
   Or, Life on the Lake.

3. ## NOW OR NEVER;
   Or, The Adventures of Bobby Bright.

4. ## TRY AGAIN;
   Or, The Trials and Triumphs of Harry West.

5. ## POOR AND PROUD;
   Or, The Fortunes of Katy Redburn.

6. ## LITTLE BY LITTLE;
   Or, The Cruise of the Flyaway.

This is the first series of books written for the young by "Oliver Optic." It laid the foundation for his fame as the first of authors in which the young delight, and gained for him the title of the Prince of Story-Tellers. The six books are varied in incident and plot, but all are entertaining and original.

THE GREAT WESTERN

SERIES.

Six Volumes. Illustrated. Per vol., $1.50.

1. GOING WEST;
   Or, The Perils of a Poor Boy.

2. OUT WEST;
   Or, Roughing it on the Great Lakes.

3. LAKE BREEZES;
   Or, The Cruise of the Sylvania.

4. GOING SOUTH;
   Or, Yachting on the Atlantic Coast.

5. DOWN SOUTH;
   Or, Yacht Adventures in Florida. (In Press.

6. UP THE RIVER;
   Or, Yachting on the Mississippi. (In Press.)

This is the latest series of books issued by this popular writer, and deals with Life on the Great Lakes, for which a careful study was made by the author in a summer tour of the immense water sources of America. The story, which carries the same hero through the six books of the series, is always entertaining, novel scenes and varied incidents giving a constantly changing, yet always attractive aspect to the narrative. "Oliver Optic" has written nothing better.

YACHT CLUB SERIES.

Uniform with the ever popular "Boat Club," Series, Completed
in six vols. 16mo. Illustrated. Per vol., $1.50.

1. **LITTLE BOBTAIL;**
   Or, The Wreck of the Penobscot.

2. **THE YACHT CLUB;**
   Or, The Young Boat-Builders.

3. **MONEY-MAKER;**
   Or, The Victory of the Basilisk.

4. **THE COMING WAVE;**
   Or, The Treasure of High Rock.

5. **THE DORCAS CLUB;**
   Or, Our Girls Afloat.

6. **OCEAN BORN;**
   Or, The Cruise of the Clubs.

The series has this peculiarity, that all of its constituent volumes are independent of one another, and therefore each story is complete in itself. "Oliver Optic" is perhaps the favorite author of the boys and girls of this country, and he seems destined to enjoy an endless popularity. He deserves his success, for he makes very interesting stories, and inculcates none but the best sentiments; and the "Yacht Club" is no exception to this rule. — *New Haven Jour. and Courier.*

THE LAKE SHORE SERIES.

Six volumes. Illustrated. In neat box. Per vol., $1.25.

1. THROUGH BY DAYLIGHT;
Or, The Young Engineer of the Lake Shore
Railroad.

2. LIGHTNING EXPRESS;
Or, The Rival Academies.

3. ON TIME;
Or, The Young Captain of the Ucayga Steamer

4. SWITCH OFF;
Or, The War of the Students.

5. BRAKE-UP;
Or, The Young Peacemakers.

6. BEAR AND FORBEAR;
Or, The Young Skipper of Lake Ucayga.

"Oliver Optic" is one of the most fascinating writers for youth, and withal one of the best to be found in this or any past age. Troops of young people hang over his vivid pages, and not one of them ever learned to be mean, ignoble, cowardly, selfish, or to yield to any vice from anything they ever read from his pen. — *Providence Press.*